Stochastic Equations for Complex Systems

Mathematics and Its Applications (*Soviet Series*)

A. V. Skorohod
Mathematics Institute, Kiev State University, U.S.S.R.

Stochastic Equations for Complex Systems

D. Reidel Publishing Company
A MEMBER OF THE KLUWER ACADEMIC PUBLISHERS GROUP
Dordrecht / Boston / Lancaster / Tokyo

Library of Congress Cataloging in Publication Data

CIP

Skorohod, A. V. (Anatoliĭ Vladimirovich), 1930–
Stochastic equations for complex systems.

(Mathematics and its applications. East European series)
Translation of: Stokhasticheskie uraveniîa dlîa slozhnykh sistem.
1. Stochastic processes. I. Title. II. Series: Mathematics and its applications (D. Reidel Publishing Company). East European series.
QA274.S58613 1987 519.2 87–20662

ISBN 978-94-010-8177-1 ISBN 978-94-009-3767-3 (eBook)
DOI 10.1007/978-94-009-3767-3

Published by D. Reidel Publishing Company,
P.O. Box 17, 3300 AA Dordrecht, Holland.

Sold and distributed in the U.S.A. and Canada
by Kluwer Academic Publishers,
101 Philip Drive, Assinippi Park, Norwell, MA 02061, U.S.A.

In all other countries, sold and distributed
by Kluwer Academic Publishers Group,
P.O. Box 322, 3300 AH Dordrecht, Holland.

This is a translation of the original work: СТОХАСТИЧЕСКИЕ УРАВНЕНИЯ ДЛЯ СЛОЖНЫХ СИСТЕМ
Published by Nauka, Moscow

Translated by L. F. Boron

TABLE OF CONTENTS

EDITOR'S PREFACE

Approach your problems from the right end and begin with the answers. Then one day, perhaps you will find the final question.

'The Hermit Clad in Crane Feathers' in R. van Gulik's *The Chinese Maze Murders*.

It isn't that they can't see the solution. It is that they can't see the problem.

G.K. Chesterton. *The Scandal of Father Brown* 'The point of a Pin'.

Growing specialization and diversification have brought a host of monographs and textbooks on increasingly specialized topics. However, the "tree" of knowledge of mathematics and related fields does not grow only by putting forth new branches. It also happens, quite often in fact, that branches which were thought to be completely disparate are suddenly seen to be related.

Further, the kind and level of sophistication of mathematics applied in various sciences has changed drastically in recent years: measure theory is used (non-trivially) in regional and theoretical economics; algebraic geometry interacts with physics; the Minkowsky lemma, coding theory and the structure of water meet one another in packing and covering theory; quantum fields, crystal defects and mathematical programming profit from homotopy theory; Lie algebras are relevant to filtering; and prediction and electrical engineering can use Stein spaces. And in addition to this there are such new emerging subdisciplines as "experimental mathematics", "CFD", "completely integrable systems", "chaos, synergetics and large-scale order", which are almost impossible to fit into the existing classification schemes. They draw upon widely different sections of mathematics. This programme, Mathematics and Its Applications, is devoted to new emerging (sub)disciplines and to such (new) interrelations as exempla gratia:

- a central concept which plays an important role in several different mathematical and/or scientific specialized areas;
- new applications of the results and ideas from one area of scientific endeavour into another;
- influences which the results, problems and concepts of one field of enquiry have and have had on the development of another.

The Mathematics and Its Applications programme tries to make available a careful selection of books which fit the philosophy outlined above. With such books, which are stimulating rather than definitive, intriguing rather than encyclopaedic, we hope to contribute something towards better communication among the practitioners in diversified fields.

Because of the wealth of scholarly research being undertaken in the Soviet Union, Eastern Europe, and Japan, it was decided to devote special attention to the work emanating from these particular regions. Thus it was decided to start three regional series under the umbrella of the main MIA programme.

The word complex in the title of this book means both large and complex. Large in the sense of stochastic systems with a large or infinite dimensional state space and complex in the sense of stochastic systems with topologically complicated state spaces. Motivations for studying such large and complex systems are many and various. There are questions of justification and derivation of macroscopic equations, there are problems of aggregation and making efficient use of as many cancellations as possible. And always there is at the back the idea that a suitable infinite (particle) limit will be simpler than the large system itself.

There is also a new tool developed here and it is the main tool: infinite systems of linear stochastic equations.

All in all, in my opinion, a fascinating and stimulating book which I accepted into this series with alacrity. Of course the reputation of the author greatly helped.

The unreasonable effectiveness of mathematics in science ...

Eugene Wigner

Well, if you know of a better 'ole, go to it.

Bruce Bairnsfather

What is now proved was once only imagined.

William Blake

As long as algebra and geometry proceeded along separate paths, their advance was slow and their applications limited.

But when these sciences joined company they drew from each other fresh vitality and thenceforward marched on at a rapid pace towards perfection.

Joseph Louis Lagrange.

Bussum, October 1986 Michiel Hazewinkel

PREFACE

It is really possible to assign an exact mathematical definition to the concept of a complex stochastic system (yes, and perhaps!)? However, there are systems whose complexity becomes evident - these are systems with an infinite-dimensional phase space. Another type of complexity is connected with the topological structure of the phase space - which also manifests itself in the finite-dimensional case. Among the complex phase spaces are those that consist of several components of different dimensions. The complex stochastic systems considered in this book are precisely systems with phase spaces of the indicated types.

Although the general theory of random processes (among them, the Markov stochastic processes) allow us to investigate processes in quite general phase spaces, the approach discussed in this book is new and, in my opinion, deserves attention.

Firstly, this approach allows us to construct processes in rather general phase spaces with the aid of stochastic differential equations (using, in this connection, the new method of investigating infinite linear systems of such equations). Secondly, it is clarified that under specified conditions a higher degree of complexity in the system (i.e. an increase in the number of interacting particles in the system) leads in the limit to an essential simplification of the system: it permits us to consider a system in X rather than systems in the phase space X^{∞}. The approach discussed for stochastic systems is also applicable to the deterministic systems considered in statistical mechanics, and allows us to obtain Boltzmann-type equations by purely mathematical means.

Finally, we must point out that the possibility of obtaining a probabilistic Brownian motion from the equations of motion of a system of particles has interested me for a long time. This interest also stimulated to a significant degree all the investigations carried out in this book. It appears to me that, to a certain extent, we have successfully clarified the situation in Brownian motion and diffusion processes.

I wish to emphasize that the book is of a purely mathematical nature - all assertions are theorems referring to mathematical objects and the explanations of a physical nature are very illustrative.

In conclusion, I wish to express my gratitude to L.V. Lobanova and N.F. Ryabova for their significant help in the technical aspects of the manuscript for this book.

Author

INTRODUCTION

1. The study of systems that are subject to random actions comprises an essential part of investigations in physics and in mechanics, and during the last decades also in cybernetics and in theoretical biology. The fundamental mathematical model for such a system is a random process and, in particular, a Markov process. The latter class of random processes was introduced by A.N. Kolmogorov [24] to describe dynamical systems that occur under the influence of random actions that are independent at various moments of time. Kolmogorov also suggested an analytic apparatus for studying Markov processes - differential equations for the probabilistic shift of the process. The idea of obtaining a Markov process from a dynamical system that occurs under the action of a process with independent values was realized by N.N. Bogolyubov and N.M. Krylov [2]. The development of the idea of Bogolyubov and Krylov led I.K. Gihman to the construction of the general theory of dynamical systems that occur under the action of a random process [9, 10] and then to the creation of the theory of stochastic equations [11, 12]. We note that at about this same time the Japanese mathematician K. Itô [20, 21, 22] was independently developing the theory of stochastic differential equations. (Although Itô started with ideas far removed from physical ones, the approach suggested by him, based on the concept of a stochastic integral, turned out also to be more suitable for describing mechanical and physical systems.) Stochastic differential equations, having arisen in the formalization of the concept of a dynamical system that occurs under a random action, became a powerful apparatus for the study not only of real systems but also of random processes.

2. In the theory of random processes, stochastic differential equations are used to effectively construct a side class of random processes, the first among these being Markov processes. In the works of K. Itô [22] diffusion processes had already been constructed on differentiable manifolds. But he also constructed one-dimensional discontinuous processes. Multidimensional discontinuous Markov processes were constructed by A.V. Skorohod [31]. In the works of V.V. Baklan [1], C.L. Cantladze [28], and Yu. L. Daleckii [5], the thoery of stochastic differential equations was extended to processes in Hilbert space, which enabled one to construct a certain class of diffusion processes in a Hilbert phase space and to carry over to a significant degree to these spaces the finite-dimensional theory. All this pertained to diffusion equations with smooth coefficients. Equations with non-smooth and continuous as well as discontinuous coefficients were begun to be studied by A.V. Skorohod [31] who used

for the proof of existence the principle of the compactness of measures, corresponding to random processes. For diffusion processes with non-smooth coefficients, the most general results were obtianed by N.V. Krylov [25, 26] and by D.V. Strook and S.R.S. Varadhan [35]. The works of N.I. Portenko [29,30] considered equations with generalized shift. At the same time, the theory of equations with coefficients depending on the entire past were worked together with representations of processes in the form of solutions of such equations. The first work in this direction is due to K. Itô and M. Nisio [23]. Problems of filtering of random processes relative to this type of equation were considered by A.N. Shiryaev [39], R.S. Lipcer and A.N. Siryaev [27] and M.P. Ersov [8]. The general theory of stochastic equations with coefficients depending on the entire past is constructed in the work of I.I. Gihman and I.I. Kadyrova [13]. The representation of discontinuous processes as solutions of stochastic equations was obtained by B. Grigelionis [13, 14].

3. The first chapter of this book is devoted to stochastic differential equations that are connected with some general class of continuous Markov processes - the so-called quasi-diffusion processes introduced by E.B. Dynkin [7]. In the work of A.V. Skorohod [32], it was shown that a rather wide class of continuous Markov processes can be transformed into a quasi-diffusion process by a random change in time. For the construction of a stochastic equation at that point, it was assumed that for a quasi-diffusion process x_t for arbitrary function φ on the domain of the definition of an infinitesimal operator of the process A, the expression

$$\varphi(x_t)-\int_0^t A\varphi(x_s)\,ds$$

is a continuous martingale with characteristic

$$\int_0^t [A\varphi^2(x_s)-2\varphi(x_s)\,A\varphi(x_s)]\,ds.$$

This situation allows us to write the following relation:

$$d\varphi(x_t)=a(\varphi,\ x_t)\,dt+(b(\varphi,\ x_t),\ dw(t))_H, \tag{1}$$

where the function $a(\varphi, x) = A\varphi(x)$, $b(\varphi, x)$ is linear in φ with values in the Hilbert space H, for which

$$(b(\varphi,\ x),\ b(\varphi,x))_H=A\varphi^2(x)-2\varphi(x)\,A\varphi(x),$$

and w(t) is a Wiener process in H with unit correlation operator.

Relation (1) can be viewed as a system of equations relative to $\varphi(x)$. This system is overdetermined, but is suitable for an arbitrary topological phase space (i.e. neither finite-dimensionality nor linearity are required). In Chapter 1 the theory of equations of type (1) is constructed for a locally compact phase space. Conditions for the existence and uniqueness of a solution are found; conditions for the weak existence and weak uniqueness are found; and transformations of such an equation that preserve its form are investigated. As examples of this type of equation we consider equations on manifolds with a boundary having a reflection on the boundary, equations on ramified manifolds, and on manifolds of variable dimension. Systems with such phase spaces have a completely reasonable physical characteristic. The presence of a random perturbation does not permit us to apply to them a method of periodic investigation for deterministic systems since a system that falls into critical states (on the boundary, on the intersection of components of different dimensions, at the branching point) will return many times (to be precise, an infinite number of times) to such states before it departs from them at some distance. The approach suggested in Chapter 1 is, at present, the only one that allows us to consider systems of this kind: namely, the complexity of the phase space compels us to treat them in the class of complex systems.

4. The second type of complex systems is considered in Chapter 2. There we consider systems consisting of a large number of randomly interacting particles, and we investigate the behavior of such systems when their number increases indefinitely. An essential difference between these systems and systems considered in statistical physics (a rather complete survey of the ideas and methods of statistical mechanics is contained in G. Uhlenbeck and J. Ford [36]) is in the randomness of the interaction, whereas in statistical physics randomness enters only through the initial state. Under the assumption that interactions between different pairs of particles are independent and the number of interactions on one particle in a unit of time remains bounded, we can investigate the limiting behavior of the system as the number of particles increases indefinitely. The equations of motion of such a system will be stochastic differential equations of the form

$$dz_i(t) = [A(z_i) + \sum_{j=1}^{n} a_n(z_i, z_j)]\,dt + \int f(\theta, z_i, z_j)\, p_{ij}^{(n)}(d\theta \times dt),$$

where z_i defines the state of the i-th particle in the phase space Z, A is the external field, $a_n(z_i, z_j)$ is the nonrandom force of interaction between the i-th and j-th particles, which can be characterized as 'remote action'. The integral with respect to the stochastic Poisson measure $p_{ij}^{(n)}$ on $[0, \infty) \times \Theta$, where Θ is some auxiliary measurable space, represents an impulse random force of interaction that changes the state of particles in a jumplike manner (the impulses of interacting particles change by jumps). The solution of the equation $(z_1(t), \ldots, z_n(t))$ will be a Markov process. The Kolmogorov equation for the distribution of the process (direct equation) plays the role of the Liouville equation. In Chapter 3, we introduce a 'statistical' distribution function:

$$\mu_t^{(n)}(A) = \frac{1}{n}\sum_{i=1}^{n} \chi_A(z_i(t))$$

(χ_A is the indicator of the Borel set A) and we study the limiting behavior of this random measure as $n \to \infty$.

We note that distributions in Z^k

$$m_t^{(k)}(dz_1, \ldots, dz_k) = \mathsf{M}\mu_t^{(n)}(dz_1) \ldots \mu_t^{(n)}(dz_k)$$

are the analogue of special distribution functions in statistical mechanics - for which we can write out a chain of equations analogous to the chain of N.N. Bogolyubov (see [3]). In our case, this has the form

$$\frac{\partial}{\partial t}\rho_t^{(k)}(z_1, \ldots, z_k) = -\sum_{r=1}^{k} \mathrm{Sp}\frac{\partial}{\partial z_r}(\rho_t^{(k)}(z_1, \ldots, z_k) A(z_r)) -$$
$$-\sum_{r=1}^{k}\int \mathrm{Sp}\frac{\partial}{\partial z_r}(\rho_t^{(k+1)}(z_1, \ldots, z_{k+1}) a(z_r, z_{k+1}))\, dz_{k+1} +$$
$$+\sum_{r=1}^{k}\int\int [\rho_t^{(k+1)}(z_1, \ldots, z_{r-1}, u, z_{r+1}, \ldots, z_{k+1}) -$$
$$-\rho_t^{(k)}(z_1, \ldots, z_k)]\,\pi(u, z_r, z_{k+1})\, du\, dz_{k+1},$$

where $\rho_t^{(k)}$ is the density of the measure $m_i^{(k)}$ relative to the Lebesgue measure in Z^k (we assume Z to be a Euclidean space), $\partial/\partial z(\rho A)$ and $\partial/\partial z(\rho a)$ are operators, whose traces appear in the equations, $\pi(u, z, \bar{z})$ is the density of the distribution of states of the particles which, before interaction, was situated at the point z and interacted with the particle situated at the point $\bar{z}$.

Let $a_n = a/n$, $Mp_{ij}^{(n)}(d\theta \times dt) = \frac{1}{n} m(d\theta)\, dt$,

where m is a finite measure on Θ. Then under the assunption of certain smoothness of the coefficients of the equations the following assertions are valid:

(1) The measure $\mu_i^{(n)}(A)$ converges weakly to some nonrandom

measure $\lambda_t(A)$ for which the following equation is satisfied

$$\frac{d}{dt}\int \lambda_t(dz)\varphi(z) = \int (\varphi'(z), A(z) + \int a(z, z')\lambda_t(dz'))\lambda_t(dz) + \\ + \int\int\int [\varphi(z + f(\theta, z, z')) - \varphi(z)]\, m(d\theta)\lambda_t(dz')\lambda_t(dz),$$

This equation can be considered as the analogue of the Boltzmann equation in statistical physics; if the density $\lambda_t(dz)$ relative to the Lebesgue measure $\rho_t(z)$ exists, this will be a limiting one-particle distribution function (in the terminology of statistical mechanics), and the equation

$$\frac{d}{dt}\rho_t(z) = -\mathrm{Sp}\frac{\partial}{\partial z}(\rho(z)A(z)) - \\ - \int \mathrm{Sp}\frac{\partial}{\partial z}(\rho_t(z)a(z, z'))\rho_t(z')dz' + \\ + \int\int [\rho_t(u)\rho_t(z') - \rho_t(z)]\pi(u, z, z')\,du\,dz'.$$

will be valid for it.

(2) Suppose the initial values of the functions $z_i^{(n)}(t)$, ..., $z_k^{(n)}(t)$ (this is the solution of equation (1) for given n) converges to $z_1(0)$, ..., $z_k(0)$. Then the joint distribution of the process $(z_1^{(n)}(t)$, ..., $z_k^{(n)}(t))$ converges as $n \to \infty$ to the joint distribution of the k independent processes $(z_1(t), \ldots, z_k(t))$, each of which is a Markov process that satisfies the stochastic differential equation

$$dz_i(t) = \hat{a}(t, z_i(t))\,dt + \int f(\theta, z_i(t), z')\,\hat{p}(d\theta \times dz' \times dt), \quad (2)$$

where $\hat{a}(t, z) = A(z) + \int a(z, z')\lambda_t(dz')$ and $\hat{p}$ is a Poisson measure on $\Theta \times Z \times [0, \infty)$. It follows from this, in particular, that the particle distribution functions are, in the limit, products of one-particle distribution functions.

(3) We consider the fluctuation of the statistical distribution function

$$\nu_t^{(n)}(A) = \sqrt{n}\,(\mu_t^{(n)}(A) - \lambda_t(A)).$$

Under certain additional assumptions the distributions of the quantities

$$\int \varphi(z)\,\nu_t^{(n)}(dz)$$

converge as $n \to \infty$ to a Gaussian distribution, i.e. $\nu_t^{(n)}(A)$ converges weakly to some generalized Gaussian field $\nu_t(dz)$. For $a(z) = A(z, z') = 0$, this field satisfies the following equation:

$$d\int \varphi(z)\nu_t(dz) = \\ = \iiint [\varphi(z + f(\theta, z, z')) - \varphi(z)]\, m(d\theta)\, [\lambda_t(dz)\nu_t(dz') + \\ + \lambda_t(dz')\nu_t(dz)] + \\ + \iiint [\varphi(z + f(\theta, z, z')) - \varphi(z)]\, \gamma(d\theta \times dz \times dz' \times dt),$$

where $\gamma(d\Theta \times dz \times dz' \times dt)$ is a Gaussian measure with independent values for which $M\gamma = 0$, $M\gamma^2(d\Theta \times dz \times dz' \times dt) = m(d\Theta)\lambda(t, dz)\lambda(t, dz')\, dt$. The solution of the preceding equation can be written in explicit form.

(4) Finally, a diffusion approximation has been obtained for equations of type (2). It is assumed that the motion is continuous; only the velocity can vary in a jumplike manner. Moreover, there exists a braking force which is directed opposite the velocity and is approximately proportional to it (viscosity). If the viscosity increases indefinitely, then the solution of equation (2) converges weakly to the diffusion process whose coefficients are expressed in terms of the coefficients of the initial equation (2). This result is analogous to the hydrodynamic approximation in statistical mechanics (see th work of N.N. Bogolyubov [4]). It also serves to clarify the fact that a probabilistic-theoretic approach makes it possible to describe the diffusion by first-order stochastic equations, and, at the same time, equations of dynamics by second-order stochastic equations. The probabilistic model of obtaining a Wiener process from a process of motion under the action of random forces was first proposed by A.M. Il'in and R.Z. Has'minskii [19]. More general classes of processes were considered in the works of A.V. Skorohod [33] and V.A. Dubko [16].

Only systems of particles of one type were considered. However, the extension of the results to systems consisting of particles of several types encounters no fundamental difficulties and can be carried out perfectly analogously.

In this book we investigate the behavior of a large number of particles only under the assumption that the particles fill a bounded set, or that a probabilistic distribution function corresponds to them. This will be the case if, at the initial moment of time $\mu_0^{(n)}(A)$ has a limit in the form of a probability measure; thus excluding, for example, the case where the particles are distributed in the limit uninformly over the entire space.

A second essential restriction is boundedness (even smoothness) of the interaction forces. The case when the forces can increase indefinitely as the particles approach each other is more interesting. Any results that would enable us to reduce the above restriction would be of considerable interest.

In this book, without special reference to basic results, we make use of the theory of martingales, the theory of stochastic differential equations, and the semigroup theory of Markov processes. These results can be found in references [14, 15, 7] and [28].

Chapter 1

CONTINUOUS MARKOV PROCESSES IN A LOCALLY COMPACT SPACE AND STOCHASTIC EQUATIONS

When considering mechanical and physical systems that occur under the action of random forces, under the assumption that the random actions are independent on nonintersecting intervals of time, it is natural to describe the variation of their states with the aid of a Markov process. If the probabilistic nature of the acting forces depends only on the state of the process and does not change with time, then the evolution of the system is described by a Markov process that is homogeneous with respect to time. In this chapter, we consider systems with compact and locally compact phase spaces. These may be certain regions of finite-dimensional Euclidean space which may be rather irregular if the systems under consideration describe the motion of several solids with complicated connections between them. The phase space can also be fundamentally infinite-dimensional, i.e. the systems can have an infinite number of degrees of freedom. The basic peculiarity of the processes being considered consists in the absence of a linear structure in the phase space. Therefore, the formulation of stochastic differential equations that describe the motion of the particles is impossible. To this end, we use a certain collection of 'pseudo-coordinates'. In order to define the possible form of stochastic equations, we first establish the form of equations for quasi-diffusion processes; then, independently, we study solutions of the equations obtained. One must note that when solving mechanical and physical problems, it is natural to start with stochastic differential equations, inasmuch as the laws of mechanics and physics are formulated in differential form. However, in practice, it is in fact not the trajectories of the system in the phase space that interest us, but only the probabilities connected with it; for example, the transition probabilities. Therefore, having constructed the solution of a stochastic differential equation, we shall further study the probabilistic characteristics of the Markov process formed by solutions of the stochastic equation.

1. Quasi-diffusion Processes

Let X be a locally compact metric space, $r(x, y)$ the distance in X, A a σ-algebra of Borel sets. We shall consider continuous homogeneous Markov processes $\{x_t\}$ in X. We denote by $P(t, x, B)$, $t \geq 0$, $x \in X$, $B \in \mathfrak{B}$, the transition probability of the process x_t from the state x into the set B at time t. Concerning the transition probability, it is assumed that the usual requirements are fulfilled: (a) $P(t, x, \cdot)$ is a probability measure on $\mathfrak{B}$, (b) $P(t, x, B)$ is measurable relative to

$\mathbf{A}_+ \times \mathbf{B}$, where $\mathbf{A}_+$ is a σ-algebra of Borel sets on $R_+ = [0, \infty)$, (c) the Chapman-Kolmogorov equation is satisfied: for $t > 0$, $s > 0$,

$$P(t+s, x, B) = \int P(t, x, dy) P(s, y, B). \tag{1}$$

We denote by C_0 the set of continuous real functions $\varphi(x)$ on X for which $\varphi(x) \to 0$ as $x \to \infty$. We shall consider only regular Feller processes, i.e. processes for which, for $\varphi \in C_0$,

$$\int \varphi(y) P(t, x, dy) \in C_0. \tag{2}$$

This condition is equivalent to the following: (1) for fixed t, the measures $P(t, x, \cdot)$ are weakly continuous with respect to x, i.e. $\int \varphi(y)P(t, x, dy)$ is continuous with respect to x for every bounded continuous function φ; processes with such transition probabilities are called Feller processes; (2) for each compactum K, $\lim_{x\to\infty} P(t, x, K) = 0$, this is the condition for the regular returnability from infinity.

REMARK. If X is a compactum, then C_0 coincides with the set C of all continuous functions, condition (2) loses its meaning and equation (2) is equivalent to the process being a Feller process.

We shall consider C_0 to be a Banach space with norm $||\varphi|| = \sup |\varphi(x)|$. We associate with a Markov process a semigroup of linear operators in C_0,

$$T_t\varphi(x) = \int \varphi(y) P(t, x, dy). \tag{3}$$

That this is a semigroup follows from (1): $T_{t+s} = T_t T_s$. Moreover, $||T|| \leq 1$ and $T_t\varphi \geq 0$ if $\varphi \geq 0$. The infinitesimal generator of the semigroup defined for all φ for which the limit

$$\lim_{t\to 0} \frac{1}{t}(T_t\varphi(x) - \varphi(x)) = A\varphi(x) \in C_0 \tag{4}$$

exists in the sense of bound pointwise convergence, is called the infinitesimal operator of the Markov process. We shall denote its domain by D. It coincides with the range of the resolvent of the semigroup (or if the Markov process) defined by the equation

$$R_\lambda f(x) = \int_0^\infty e^{-\lambda t} T_t f(x)\, dt, \qquad f \in C_0. \tag{5}$$

In this connection

$$AR_\lambda f = \lambda R_\lambda f - f, \quad f \in C_0; \; R_\lambda A f = \lambda R_\lambda f - f, \quad f \in D,$$

and hence $R_\lambda = (\lambda I - A)^{-1}$, where I is the identity operator in C_0. Using (5), one can reconstruct the semigroup via the resolvent and, hence, also via the infinitesimal generator. If $\varphi \in D$, then $T_t\varphi \in D$ for $t > 0$ and

$$\frac{d}{dt} T_t\varphi(x) = T_t\varphi(x). \tag{6}$$

From this there follows the relation

$$T_t\varphi(x) - \varphi(x) = \int_0^t T_s A\varphi(x)\, ds. \tag{7}$$

The preceding equality allows the following interesting interpretation. We shall denote by P_x the probability calculated under the condition that the initial state of the process x_t coincides with x ($x_0 = x$); M_x is the mathematical expectation with respect to the probability P_x. Then

$$\mathsf{M}_x\varphi(x_t) = T_t\varphi(x), \; \mathsf{M}_x\varphi(x_t) - \varphi(x) = \mathsf{M}_x \int_0^t A\varphi(x_s)\, ds,$$

or

$$\mathsf{M}_x\left[\varphi(x_t) - \varphi(x_0) - \int_0^t A\varphi(x_s)\, ds\right] = 0.$$

Let F_t be the filtration of the σ-algebra generated by the process x_t. We shall show that the process

$$\zeta_t = \varphi(x_t) - \int_0^t A\varphi(x_s)\, ds$$

is a martingale relative to F_t with respect to the measure P_x, for arbitrary $x \in X$. In fact,

$$= \mathsf{M}_x\left[\varphi(x_{t+h}) - \varphi(x_t) - \int_t^{t+h} A\varphi(x_s)\, ds / \mathcal{F}_t\right] =$$

$$= \mathsf{M}\left[\varphi(x_{t+h}) - \varphi(x_t) - \int_t^{t+h} A\varphi(x_s)\,ds / x_t\right] =$$

$$= \mathsf{M}_{x_t}\left[\varphi(x_h) - \varphi(x_0) - \int_0^h A\varphi(x_s)\,ds\right] = 0;$$

we made use of the Markov property and of the homogeneity of the process. We shall say that φ belongs to the domain $\tilde{D}$ of the quasi-infinitesimal operator $\tilde{A}$ if there exists a bounded measurable function g(x) such that the process

$$\varphi(x_t) - \int_0^t g(x_s)\,ds$$

is a martingale relative to F_t with respect to the measure P_x, for arbitrary $x \in X$. In this case, the operator $\tilde{A}$ is defined on φ and $\tilde{A}\varphi = g$. It follows from what we states above that $\tilde{A}$ is an extension of the operator A.

We now recall the definition of a diffusion process. A continuous Markov process x_t is called a diffusion process if X is a finite-dimensional Euclidean space and every twice continuously differentiable function having bounded derivatives belong to $\tilde{D}$. (This is somewhat wider than the conventional definition.) In particular, if $X = R^m$, x^i are the coordinates of the point x, and $g_c(x)$ is a twice continuously differentiable function, equal to 1 for $|x| < c$ and tending to zero together with its derivatives as $x \to \infty$, then the functions $x^i g_c(x)$, $x^i x^j g_c(x)$ belong to $\tilde{D}$. For $|x| < c$, the quantities

$$a^i(x) = \tilde{A}(x^i g_c(x)),\quad b^{ij}(x) = \tilde{A}(x^i x^j g_c(x))$$

do not depend on c. They are called diffusion coefficients. Through these coefficients the operator A is defined on twice differentiable functions with the aid of the formula

$$\tilde{A}\varphi(x) = \sum_{i=1}^m a^i(x)\frac{\partial\varphi(x)}{\partial x^i} + \frac{1}{2}\sum_{i,j=1}^m b^{ij}(x)\frac{\partial^2\varphi(x)}{\partial x^i \partial x^j}. \tag{8}$$

We note that for a diffusion process in the generally accepted terminology the coefficients $a^i(x)$ and $b^{ij}(x)$ must satisfy the additional condition

$$\lim_{s\to 0}\frac{1}{s}\int_0^s \mathsf{M}_x(|a^i(x)-a^i(x_t)|+|b^{ij}(x)-b^{ij}(x_t)|)\,dt=0.$$

A diffusion process possesses the following property: twice continuously differentiable functions that coincide locally with x^i belong to $\tilde{D}$ and every twice continuously differentiable function with bounded derivatives of $x^1, \ldots, x^m$ belongs to $\tilde{D}$. And a diffusion process is defined starting with this property.

DEFINITION. A continuous Markov process x_t is called a quasi-diffusion process if, for arbitrary $\varphi_1, \varphi_2, \ldots, \varphi_n \in \tilde{D}$ and a twice continuously differentiable function of n variables $F(\xi_1, \ldots, \xi_n)$, the function $F(\varphi_1(x), \ldots, \varphi_n(x))$ belongs to $\tilde{D}$.

In particular, for a quasi-diffusion process, it follows from $\varphi \in \tilde{D}$ that the function φ^2 also belongs to $\tilde{D}$.

We recall that a nondecreasing continuous function $\langle\zeta\rangle_t$, adapted to the filtration F_t such that $\zeta_t^2 - \langle\zeta\rangle_t$ is also a martingale, is called the (quadratic) characteristic of a continuous martingale ζ_t. The property defining a quasi-diffusion process makes it possible to calculate the characteristic of the martingale

$$\zeta_t=\varphi(x_t)-\int_0^t \tilde{A}\varphi(x_s)\,ds \tag{9}$$

for $\varphi \in \tilde{D}$. In fact

$$\zeta_t^2=\varphi^2(x_t)-2\varphi(x_t)\int_0^t \tilde{A}\varphi(x_s)\,ds+\left(\int_0^t \tilde{A}\varphi(x_s)\,ds\right)^2=$$
$$=\varphi^2(x_t)-2\zeta_t\int_0^t \tilde{A}\varphi(x_s)\,ds-\left(\int_0^t \tilde{A}\varphi(x_s)\,ds\right)^2.$$

Using the differentiability of the integral of $\tilde{A}\varphi$, we can convince ourselves (in particular, this follows from the Itô formula for martingales) that

$$d\left[2\zeta_t\int_0^t \tilde{A}\varphi(x_s)\,ds+\left(\int_0^t \tilde{A}\varphi(x_s)\,ds\right)^2\right]=$$
$$=2\left[\zeta_t\tilde{A}\varphi(x_t)\,dt+\tilde{A}\varphi(x_t)\int_0^t \tilde{A}\varphi(x_s)\,ds\,dt+\int_0^t \tilde{A}\varphi(x_s)\,ds\,d\zeta_t\right]=$$
$$=2\varphi(x_t)\,\tilde{A}\varphi(x_t)\,dt+2\left(\int_0^t \tilde{A}\varphi(x_s)\,ds\right)d\zeta_t.$$

Consequently

$$\zeta_t^2 = \varphi^2(x_t) - \int_0^t \tilde{A}\varphi^2(x_s)\,ds - 2\int_0^t \left(\int_0^s \tilde{A}\varphi(x_u)\,du\right) d\zeta_s +$$

$$+ \int_0^t [\tilde{A}\varphi^2(x_s) - 2\varphi(x_s)\tilde{A}\varphi(x_s)]\,ds.$$

Inasmuch as $\varphi^2(x_t) - \int_0^t A\varphi^2(x_s)\,ds$ is a martingale by the definition of $\tilde{A}$,

$$\int_0^t \left(\int_0^s \tilde{A}\varphi(x_u)\,du\right) d\zeta_s$$

is a martingale, being the stochastic integral with respect to a martingale, then

$$\zeta_t^2 - \int_0^t [\tilde{A}\varphi^2(x_s) - 2\varphi(x_s)\tilde{A}\varphi(x_s)]\,ds$$

is a martingale. Hence

$$\langle\zeta\rangle_t = \int_0^t [\tilde{A}\varphi^2(x_s) - 2\varphi(x_s)\tilde{A}\varphi(x_s)]\,ds. \qquad (10)$$

Thus, the characteristic of the martingale (9) is defined by formula (10). In particular, it is an absolutely continuous function of t and its derivative with respect to t is bounded.
      We shall prove the converse: if $\langle h\rangle_t = \int_0^t g(s)\,ds$, where $g(s)$ is F_s-adapted, measurable and bounded, then $\varphi^2 \in D$. The equation

$$\zeta_t^2 = \varphi^2(x_t) - 2\int_0^t \left(\int_0^s \tilde{A}\varphi(x_u)\,du\right) d\zeta_s - \int_0^t 2\varphi(x_s)\tilde{A}\varphi(x_s)\,ds$$

follows from the preceding discussion. Since the characteristic is defined uniquely by the process,

$$\varphi^2(x_t) - \int_0^t [g(s) + 2\varphi(x_s)\tilde{A}\varphi(x_s)]\,ds$$

is a martingale. We note now that $\langle\zeta\rangle_t$ is a continuous homogeneous additive function of the Markov process x_t. If θ_h is the shift operator on the trajectory of the process: $\theta_h x_t = x_{t+h}$, then

$$\langle\zeta\rangle_{t+h}=\langle\zeta\rangle_h+\theta_h\langle\zeta\rangle_t \tag{11}$$

with probability $P_x = 1$, for arbitrary x. This property follows from the relation

$$\langle\zeta\rangle_t=\lim_{\max\Delta t_k\to 0}\sum_{k=0}^{n-1}(\zeta_{t_{k+1}}-\zeta_{t_k})^2,\quad 0=t_0<t_1<\ldots<t_n=t$$

(the limit is in the sense of root-mean-square convergence with respect to the measure P_x). It follows from formula (11) that

$$\int_h^{t+h} g(s)\,ds=\int_0^t \theta_h g(s)\,ds=\int_0^t g(s+h)\,ds.$$

Using the situation that for almost all s

$$g(s)=\hat{g}(s)=\overline{\lim_{\delta\downarrow 0}}\frac{1}{\delta}\int_s^{s+\delta} g(u)\,du$$

and the quantity on the right is F_s-measurable, we can assume that $\theta_h g(s) = g(s+h)$, inasmuch as this is true for $\hat{g}$. We set

$$\overline{\lim_{\delta\downarrow 0}}\frac{1}{\delta}\int_0^{\delta} g(u)\,du=f(x);$$

inasmuch as this quantity is F_0-measurable and hence constant almost everywhere with respect to the measure P_x, it can depend on x, but is a Borel function of x, coinciding with its mathematical expectation. Then $g(0) = f(x_0)$ and so $g(s) = f(x_s)$ for almost all s. Thus, the process

$$\varphi^2(x_t)-\int_0^t [f(x_s)+2\varphi(x_s)\tilde{A}\varphi(x_s)]\,ds,$$

is a martingale, whence it follows that $\varphi^2 \in \tilde{D}$.

We shall now show that if also $\varphi^2 \in \tilde{D}$ for all $\varphi \in \tilde{D}$ then the process is a quasi-diffusion process. Let $\varphi_k \in \tilde{D}$, $k = 1, \ldots, n$, and let $F(\xi_1, \ldots, \xi_n)$ be a twice continuously differentiable function in R^n. We set

$$\zeta_k(t)=\varphi_k(x_t)-\int_0^t \tilde{A}\varphi_k(x_s)\,ds.$$

Then, if $\langle\zeta_k, \zeta_j\rangle_t$ is the joint characteristic of the martingales ζ_k and ζ_j, i.e.

$$\langle\zeta_k, \zeta_j\rangle_t = \frac{1}{2}\left[\langle\zeta_k, \zeta_j\rangle_t - \langle\zeta_k\rangle_t - \langle\zeta_j\rangle_t\right],$$

then by virtue of formula (10)

$$\langle\zeta_k, \zeta_j\rangle_t = \int_0^t [\tilde{A}\varphi_k(x_s)\varphi_j(x_s) - \varphi_k(x_s)\tilde{A}\varphi_j(x_s) - \varphi_j(x_s)\tilde{A}\varphi_k(x_s)]\,ds.$$

We have

$$F(\varphi_1(x_t), \ldots, \varphi_n(x_t)) = F\left(\zeta_1(t) + \int_0^t \tilde{A}\varphi_1(x_s)\,ds, \ldots, \zeta_n(t) + \int_0^t \tilde{A}\varphi_n(x_s)\,ds\right).$$

On the basis of Itô's formula (see [15, Vol. 3, p.96]), we can write

$$F(\varphi_1(x_t), \ldots, \varphi_n(x_t)) = F(\varphi_1(x_0), \ldots, \varphi_n(x_0)) + \sum_{i=1}^n \int_0^t \frac{\partial F}{\partial \xi_i}\left(\zeta_1(s) + \int_0^s \tilde{A}\varphi_1(x_u)\,du, \ldots\right) d\zeta_i(s) + \int_0^t \left[\sum_{i=1}^n \frac{\partial F}{\partial \xi_i}\left(\zeta_1(s) + \int_0^s \tilde{A}\varphi_1(x_u)\,du, \ldots\right)\tilde{A}\varphi_i(x_s) + \frac{1}{2}\sum_{i,j}\frac{\partial^2 F}{\partial \xi_i \partial \xi_j}\left(\zeta_1(s) + \int_0^s \tilde{A}\varphi_1(x_u)\,du, \ldots\right)\frac{d}{ds}\langle\zeta_i, \zeta_j\rangle_s\right] ds.$$

Consequently, $F(\varphi_1, \ldots, \varphi_n) \in \tilde{D}$ and

$$\tilde{A}F(\varphi_1, \ldots, \varphi_n) = \sum_{i=1}^n \frac{\partial F}{\partial \varphi_i}\tilde{A}\varphi_i + \frac{1}{2}\sum_{i,j=1}^n \frac{\partial^2 F}{\partial \varphi_i \partial \varphi_j}[\tilde{A}\varphi_i\varphi_j - \varphi_i\tilde{A}\varphi_j - \varphi_j\tilde{A}\varphi_i]. \tag{12}$$

It has been proved by the same token that the process is a quasi-diffusion process.

Thus, one can still get two equivalent definitions of a quasi-diffusion process: (1) a process is a quasi-diffusion process if for all $\varphi \in \tilde{D}$ we have $\varphi^2 \in \tilde{D}$, i.e. $\tilde{D}$ is an algebra since $\tilde{D}$ is obviously linear and for $\varphi_1, \varphi_2 \in \tilde{D}$,

$$\varphi_1\varphi_2 = \frac{1}{2}(\varphi_1 + \varphi_2)^2 - \frac{1}{2}\varphi_1^2 - \frac{1}{2}\varphi_2^2 \in \tilde{D};$$

(2) a process is a quasi-diffusion process if for all $\varphi \in \tilde{D}$ the martingale $\varphi(x_t) - \int_0^t \tilde{A}\varphi(x_s)ds$ has an absolutely continuous characteristic whose derivative is bounded.

2. Stochastic Equation for Quasi-diffusion Processes

Let x_t be a quasi-diffusion process with quasi-infinitesimal operator $\tilde{A}$, having domain $\tilde{D}$. For $\varphi \in \tilde{D}$, we denote the martingale

$$\zeta_t(\varphi) = \varphi(x_t) - \varphi(x_0) - \int_0^t \tilde{A}\varphi(x_s)\,ds \tag{1}$$

by $\zeta_t(\varphi)$. Then $\varphi(x_t)$ satisfies the stochastic differential relation

$$d\varphi(x_t) = \tilde{A}\varphi(x_t)\,dt + d\zeta_t(\varphi). \tag{2}$$

In order to obtain the stochastic differential equation from this, it is necessary to express $d\zeta_t(\varphi)$ in terms of a martingale which would depend neither on φ nor on the process x_t. As an instance of such a martingale, one can take a Wiener process. To this end, we need a theorem on the representation of a family of continuous martingales with the aid of a Wiener process.

THEOREM 1. Let Φ be a separable linear space and suppose, for all $\varphi \in \Phi$, there are defined continuous martingales $\zeta_t(\varphi)$ ($\zeta_0(\varphi) = 0$) on one and the same probability space $\{\Omega, \mathfrak{S}, P\}$ the adapted to the filtration $F_t \subset \mathfrak{S}$. We assume that the joint characteristic of the martingales $\zeta_t(\varphi)$ and $\zeta_t(\psi)$ is representable in the form

$$c_t(\varphi, \psi) = \int_0^t \gamma_s(\varphi, \psi)\,ds,$$

where $\gamma_s(\varphi, \psi)$ is a measurable F_s-adapted bounded process (for given φ, ψ) and, moreover, for each t, $c_t(\varphi, \psi)$ is a bilinear form on Φ^2, continuous in the sense of convergence in probability. Then in the separable Hilbert space H one can select a Wiener process w(t) with unit correlation operator, defined on some extension of the initial probability space and adapted to some flow $\tilde{F}_t$ such that $F_t \subset \tilde{F}_t$ and an $\tilde{F}_t$-adapted measurable random function $b_t(\varphi)$ from $R_+ \times \Phi$ into H, measurable relative to $\mathfrak{A}_+ \times \mathfrak{B}_\Phi$, where $\mathfrak{B}_\Phi$ is a σ-algebra of Borel sets in Φ, such that the representation

$$\zeta_t(\varphi) = \int_0^t (b_s(\varphi), dw(s))_H \tag{4}$$

holds; $(\cdot, \cdot)_H$ is the inner product in H.

Proof. We shall need the following factored representation: there exists a measurable function $b_s(\varphi)$ from $R_+ \times \Phi$ into H, measurable relative to $\mathfrak{A}_+ \times \mathfrak{B}_\Phi$ and F_s-concordant, such that

$$c_t(\varphi, \psi) = \int_0^t (b_s(\varphi), b_s(\psi))_H \, ds. \tag{5}$$

We choose in Φ a countable dense set $\Phi_0 = \{\varphi_1, \varphi_2, \ldots\}$. We set

$$\alpha_{ik}(s) = \gamma_s(\varphi_i, \varphi_k).$$

For almost all s the matrix $||\alpha_{ik}(s)||$ is nonnegative definite. On some other probability space $\{\Theta, \mathfrak{C}, \Pi\}$ we consider the numerical measurable random functions $\xi_i(s, \omega)$ for which, for almost all s, ω $\{\xi_i(s, \omega) = \xi_i(s, \omega, \theta)\}$ is a Gaussian sequence and $M_\Pi \xi(s, \omega)\xi_k(s, \omega) = \alpha_{ik}(s, \omega)$ (M_Π is the mathematical expectation with respect to the measure Π). The existence of such processes follows from Theorem 2, given below. By virtue of the measurability of the function $\xi_i(s, \omega)$, the Hilbert space H^0 generated by the quantities ($M_\Pi \xi\eta$ is taken as the inner product of ξ and η) will be separable. Therefore, setting $b_s^0(\varphi) = \xi_i(s)$, $\varphi \in \Phi_0$, $\varphi = \varphi_i$, we shall have

$$\gamma_s(\varphi, \psi) = (b_s^0(\varphi), b_s^0(\psi))_{H^0}, \qquad \varphi, \psi \in \Phi_0,$$

$$c_t(\varphi, \psi) = \int_0^t (b_s^0(\varphi), b_s^0(\psi))_{H^0} \, ds, \quad \varphi, \psi \in \Phi_0.$$

Using the continuity of $c_t(\varphi, \psi)$ with respect to φ and ψ with respect to probability P, we convince ourselves that, as $\varphi_{n_k} \to \varphi \in \Phi$,

$$c_t(\varphi_{n_k} - \varphi_{n_i}, \varphi_{n_k} - \varphi_{n_i}) = c_t(\varphi_{n_k}, \varphi_{n_k}) + c_t(\varphi_{n_i}, \varphi_{n_i}) - 2c_t(\varphi_{n_i}, \varphi_{n_k}) \to 0$$

and so

$$\lim_{i,k\to\infty}\int_0^t |b_s^0(\varphi_{n_i})-b_s^0(\varphi_{n_k})|_{H_0}\,ds=0$$

Therefore, the function $b_s^0(\varphi)$ can be extended on Φ whereby the equation

$$c_t(\varphi,\psi)=\int_0^t (b_s^0(\varphi),\ b_s^0(\psi))_{H_0}\,ds$$

will be fulfilled. It remains to make use of the isometry of all separable Hilbert spaces and set $b_s(\varphi) = Sb_s^0(\varphi)$, where S is an isometric mapping of H^0 into H. This completes the proof of relation (6).

We denote by Q_s the operator of projection on the closure of the range of the function $b_s(\varphi)$, $\varphi \in \Phi$. We define the martingales $\eta_t(h)$, $h \in H$, adapted to the filtration F_t in the following way. Let $\alpha_k^{(n)}$ be F_s-adapted measurable functions such that for almost all s

$$|Q_s h-\sum \alpha_k^{(n)}(s)\,b_s(\varphi_k)|_H \leqslant 1/n.$$

We set

$$\eta_t^{(n)}(h)=\sum_k \int_0^t \alpha_k^{(n)}(s)\,d\zeta_s(\varphi_k).$$

Then the characteristic of the martingale $\eta_t^{(n)}(h) - \eta_t^{(m)}(h)$ will be

$$\sum_{k,j}\int_0^t [\alpha_k^{(n)}(s)-\alpha_k^{(m)}(s)][\alpha_j^{(n)}(s)-\alpha_j^{(m)}(s)]\,d\langle\zeta(\varphi_k),\ \zeta(\varphi_j)\rangle_s=$$

$$=\int_0^t\sum_{k,j}[\alpha_k^{(n)}(s)-\alpha_k^{(m)}(s)][\alpha_j^{(n)}(s)-\alpha_j^{(m)}(s)]\,(b_s(\varphi_k),\ b_s(\varphi_j))_H\,ds=$$

$$=\int_0^t\left|\sum_k \alpha_k^{(n)}(s)\,b_s(\varphi_k)\ -\sum_j \alpha_j^{(m)}(s)\,b_s(\varphi_j)\right|_H^2 ds\leqslant\frac{2}{n^2}+\frac{2}{m^2}.$$

Inasmuch as this expression tends to zero as n, m $\to\infty$, then $\eta_t^{(m)}(h)$ converges in probability to a martingale which will be continuous and its characteristic will be the limit of the characteristic of the martingale $\eta_t^{(n)}(h)$. We denote the limit martingale by $\eta_t(h)$. It is clear from the preceding calculations that the characteristic of the martingale will be

$$\int_0^t \Big| \sum_k \alpha_k^{(n)}(s)\, b_s(\varphi_k) \Big|_H^2 ds.$$

Passing to the limit, we find

$$\langle \eta(h) \rangle_t = \int_0^t |Q_s h|_H^2 ds.$$

These same considerations show that the joint characteristic of the martingales $\eta_t(h_1)$ and $\eta_t(h_2)$ will be

$$\langle \eta(h_1),\ \eta(h_2) \rangle_t = \int_0^t (Q_s h_1,\ h_2)_H\, ds.$$

Let $w_1(t)$ be a Wiener process in H with identity correlation operator that does not depend on the flow F_t. The martingale

$$\int_0^t (h - Q_s h,\ dw_1(s))_H$$

is orthogonal to any martingale $\eta_t(\varphi)$ and hence to the martingale $\eta_t(h)$. Therefore, the martingale

$$\hat{\eta}_t(h) = \eta_t(h) + \int_0^t (h - Q_s h,\ dw_1(s))_H$$

relative to the filtration $\hat{F}_t$, generated by F_t and with values $w_1(s)$, $s \leq t$, will have as characteristic the sum of the characteristics of the terms, i.e.

$$\langle \hat{\eta}(h) \rangle_t = \int_0^t |Q_s h|_H^2 ds + \int_0^t |h - Q_s h|_H^2 ds = t\,|h|_H^2.$$

Thus, for each $h \in H$, $\hat{\eta}_t(h)$ is a Wiener process (it follows from the well-known Levi theorem that a continuous martingale with nonrandom characteristic, proportional to t, is a Wiener process).

The joint characteristic of the martingales $\hat{\eta}_t(h_1)$ and $\hat{\eta}_t(h_2)$ will be $(h_1, h_2)_H$. Using this situation, we convince ourselves that

$$\langle \hat{\eta}(\alpha h) - \alpha \hat{\eta}(h) \rangle_t = 0, \quad \langle \hat{\eta}(h_1 + h_2) - \hat{\eta}(h_1) - \hat{\eta}(h_2) \rangle = 0,$$
$$\alpha \in R, \quad h,\ h_1,\ h_2 \in H.$$

For example, the second equation follows from the fact that

$$\langle\hat{\eta}(h_1+h_2)-\hat{\eta}(h_1)-\hat{\eta}(h_2)\rangle_t = \\ = t|h_1+h_2|_H^2+t|h_1|_H^2+t|h_2|_H^2-2t(h_1+h_2,\ h_1)_H - \\ -2t(h_1+h_2,\ h_2)_H+2t(h_1,\ h_2)_H=0.$$

Consequently, with probability 1,

$$\hat{\eta}_t(\alpha_1 h_1+\alpha_2 h_2)=\alpha_1\hat{\eta}_t(h_1)+\alpha_2\hat{\eta}_t(h_2),\quad \alpha_1,\ \alpha_2\in R,\quad h_1,\ h_2\in H.$$

Thus, $\hat{\eta}_t(h)$ is a linear function of h, and for each h it is a Wiener process with mean 0 and dispersion $t|h|_H^2$. By the definition of a Wiener process in H with identity correlation operation,

$$\hat{\eta}_t(h)=(h,\ w(t))_H,$$

where w(t) is such a process.

We shall show that with such a choice of w(t) equation (3) is satisfied. To this end, we find the joint characteristic of the martingales $\hat{\eta}_t(\varphi)$ and $\zeta_t(\varphi)$. Inasmuch as $\zeta_t(\varphi)$ is orthogonal to $\hat{\eta}_t(\varphi)$ - $\eta_t(h)$ for all φ, then

$$\langle\hat{\eta}(h),\ \zeta(\varphi)\rangle_t=\langle\eta(h),\ \zeta(\varphi)\rangle_t=\lim_{n\to\infty}\langle\eta^{(n)}(h),\ \zeta(\varphi)\rangle_t= \\ =\lim_{n\to\infty}\int_0^t\sum_k\alpha_k^{(n)}(s)(b_s(\varphi_k),\ b_s(\varphi))_H\,ds=\int_0^t(h,\ b_s(\varphi))_H\,ds,$$

since

$$\lim_{n\to\infty}\left|\int_0^t\Big(\sum_k\alpha_k^{(n)}(s)\,b_s(\varphi_k)-h,\ b_s(\varphi)\Big)_H ds\right|\leqslant \\ \leqslant\lim_{n\to\infty}\left(\int_0^t\left|h-\sum_k\alpha_k^{(n)}(s)b_s(\varphi_k)\right|_H^2 ds\right)^{1/2}\left(\int_0^t|b_s(\varphi)|_H^2\,ds\right)^{1/2}$$

So,

$$\langle\hat{\eta}(h),\ \zeta(\varphi)\rangle_t=\int_0^t(h,\ b_s(\varphi))_H\,ds. \tag{5}$$

If $h_t(\omega)$ is an arbitrary step function with values in H, adapted to $\hat{F}_t$, then $\int_0^t(h_s(\omega),\ dw(s))_H$ is a martingale with characteristic $\int_0^t|h_s(\omega)|^2 ds$, and the joint characteristic of this martingale with $\zeta_t(\varphi)$ will be

$$\int_0^t (h_s(\omega),\ b_s(\varphi))_H\, ds.$$

In order to convince oneself of this, it suffices to make use of formula (5) on the intervals of constancy of $h_t(\omega)$. (It is easy to see that $\hat{\eta}_t(h)$ is a martingale for $t > s$, if $h = h(\omega)$ is F_s-measurable and the joint characteristic of the martingales $\zeta_t(\varphi)$ and $\hat{\eta}_t(h(\omega))$ on $[s, \infty)$ is given by the expression $\int_s^t (h(\omega),\ b_u(\varphi))_H du$.) With the aid of passage to the limit we convince ourselves that

$$\left\langle \int_0^{\cdot} (h_s(\omega),\ dw(s))_H,\ \zeta(\varphi) \right\rangle_t = \int_0^t (h_s(\omega),\ b_s(\varphi))_H\, ds.$$

We calculate the characteristic of the difference

$$\zeta_t(\varphi) - \int_0^t (b_s(\varphi),\ dw(s))_H.$$

It equals

$$\langle \zeta(\varphi) \rangle_t + \left\langle \int_0^{\cdot} (b_s(\varphi),\ dw(s))_H \right\rangle_t - 2\left\langle \zeta(\varphi),\ \int_0^{\cdot} (b_s(\varphi),\ dw(s))_H \right\rangle_t =$$

$$= c_t(\varphi,\ \varphi) + \int_0^t (b_s(\varphi),\ b_s(\varphi))_H\, ds - 2\int_0^t (b_s(\varphi),\ b_s(\varphi))_H\, ds = 0.$$

This establishes formula (3) and the theorem is proven.

We shall apply the theorem just proved to the family of martingales $\zeta_t(\varphi)$ defined by equation (1). We shall consider this family for some fixed measure $P_{\bar{x}}$. The linear space Φ coincides with $\tilde{D}$. We shall introduce a topology in $\tilde{D}$ so that the conditions of Theorem 1 are satisfied. In our case, the function c_t has the form

$$c_t(\varphi,\ \psi) = \int_0^t [\tilde{A}\varphi(x_s)\psi(x_s) - \varphi(x_s)\tilde{A}\psi(x_s) - \psi(x_s)\tilde{A}\varphi(x_s)]\, ds. \qquad (6)$$

(this follows from formula (10), §1). We define in $\tilde{D}$ the distnace between φ and ψ with the aid of the formula

$$\rho(\varphi,\ \psi) = \Big[\sup_{t \geq 0} e^{-t} \mathsf{M}_x |\varphi(x_t) - \psi(x_t)|^2 +$$

$$+ \mathsf{M}_x \int_0^\infty (\tilde{A}\varphi(x_t) - \tilde{A}\psi(x_t))^2 e^{-t} dt\Big]^{1/2}$$

If $\rho(\varphi, \varphi_n) \to 0$, then

$$\mathsf{M}_x \sup_{t \leqslant T} |\zeta_t(\varphi) - \zeta_t(\varphi_n)|^2 \leqslant 4\mathsf{M}_x |\zeta_t(\varphi) - \zeta_t(\varphi_n)|^2 \leqslant$$
$$\leqslant 8\mathsf{M}_x |\varphi(x_T) - \varphi_n(x_T)|^2 + 8\mathsf{M}_x T \int_0^T |\tilde{A}\varphi(x_s) - \tilde{A}\varphi_n(x_s)|^2 ds \to 0.$$

From this it follows that, for all T, $\langle\zeta(\varphi) - \zeta(\varphi_n)\rangle_T \to 0$ in probability, and so $c_T(\varphi - \varphi_n, \varphi - \varphi_n) \to 0$ in probability. We now make use of the inequality

$$|c_t(\varphi, \psi) - c_t(\varphi_n, \psi_n)| = |c_t(\varphi, \psi - \psi_n) + c_t(\varphi - \varphi_n, \psi) -$$
$$- c_t(\varphi - \varphi_n, \psi - \psi_n)| \leqslant [c_t(\varphi, \varphi) c_t(\psi - \psi_n, \psi - \psi_n)]^{1/2} +$$
$$+ [c_t(\psi, \psi) c_t(\varphi - \varphi_n, \varphi - \varphi_n)]^{1/2} +$$
$$+ [c_t(\varphi - \varphi_n, \varphi - \varphi_n) c_t(\psi - \psi_n, \psi - \psi_n)]^{1/2}.$$

If $\rho(\varphi, \varphi_n) \to 0$, $\rho(\psi, \psi_n) \to 0$, then $|c_t(\varphi, \psi) - c_t(\varphi_n, \psi_n)| \to 0$ in probability.

In order to prove the separability of $\tilde{D}$ we note beforehand that the space C_0 with the metric

$$\rho_1(\varphi, \psi) = [\sup_t \mathsf{M}_x |\varphi(x_t) - \psi(x_t)|^2 e^{-t}]^{1/2},$$

is separable inasmuch as $\rho_1(\varphi, \psi) \leq ||\varphi - \psi||$, and also the space $\mathfrak{M}$ of all bounded Borel functions with metric

$$\rho_2(g_1, g_2) = \left[\mathsf{M}_x \int_0^\infty (g_1(x_t) - g_2(x_t))^2 e^{-t} dt\right]^{1/2}.$$

is separable. Consequently, the Cartesian product of these spaces, $C_0 \times \mathfrak{M}$, equipped with the metric

$$r((\varphi; g), (\varphi_1, g_1)) = \sqrt{\rho_1^2(\varphi, \varphi_1) + \rho_2^2(g_1, g_2)}.$$

will also be separable. It follows from the definitions of the metrics ρ and r that the correspondence $\varphi \to (\varphi; \tilde{A}\varphi)$ is an isometric mapping of $\tilde{D}$ into $C_0 \times \mathfrak{M}$, from which we also obtain separability of $\tilde{D}$ with the metric ρ. Thus, the conditions of Theorem 1 are satsified, and, in addition,

$$\gamma_s(\varphi, \psi) = \tilde{A}[\varphi(x_s)\psi(x_s)] - \varphi(x_s)\tilde{A}\psi(x_s) - \psi(x_s)\tilde{A}\varphi(x_s).$$

Therefore, the Gaussian random functions $\xi_i(s, \delta)$ made use of in the proof of Theorem 1, will have the form $\xi_i(x_s)$, so that the function $b_s(\varphi)$ will also have the form $b(\varphi, x_s)$, where $b(\varphi, x)$ is a $\mathcal{B}_\Phi \times \mathcal{B}$-measurable function from $\Phi \times X$ into H, and in addition

$$(b(\varphi, x), b(\varphi, x))_H = \tilde{A}\varphi^2(x) - 2\varphi(x)\tilde{A}\varphi(x) \tag{7}$$

for almost all x with respect to the measure $P(t, \bar{x}, dx)$ (this is the transition probability of the process) for almost all t. It follows from relation (7) that the equation

$$\begin{aligned}(b(\varphi, x), b(\psi, x))_H &= \\ &= \tilde{A}[\varphi(x_s), \psi(x_s)] - \varphi(x_s)\tilde{A}\psi(x_s) - \psi(x_s)\tilde{A}\varphi(x_s),\end{aligned}$$

is satisfied for these same x, whence it follows that $b(\varphi, x)$ depends 'almost linearly' on φ: for arbitrary $\alpha_0, \alpha_1 \in R$ and $\varphi, \varphi_1 \subset \tilde{D}$ the relation

$$b(\alpha\varphi + \alpha_1\varphi_1, x) = \alpha b(\varphi, x) + \alpha_1 b(\varphi_1, x)$$

is satisfied for almost all x with respect to the measure P(t, x, dx) for almost all t. Thus, equation (2) can be written in the form

$$d\varphi(x_t) = \tilde{A}\varphi(x_t)\,dt + (b(\varphi, x_t), dw(t))_H. \tag{8}$$

We denote $\tilde{A}\varphi(x) = a(\varphi, x)$. This function is defined on $\tilde{D} \times X$ and is linear with respect to φ. Making use of this notation, we can rewrite the equation in this form:

$$d\varphi(x_t) = a(\varphi, x_t)\,dt + (b(\varphi, x_t), dw(t))_H, \tag{9}$$

where the coefficients of the equation are defined on $\tilde{D} \times X$, measurable with respect to x, linear (in the sense indicated above) with respect to φ, continuous with respect to φ in the metric ρ, $\underline{a}$ takes on values from R, $\underline{b}$ from H. Equation (9) is solved with the initial condition $\varphi(x(0)) \equiv \varphi(x)$. If one succeeds in defining $\varphi(x_t)$ from this equation for φ from some set $D_0 \subset \tilde{D}$ that separates points of X, then by the same token one also succeeds in defining x_t.

We shall show that the process w(t) depends in an essential way on the value $\bar{x}$ inasmuch as it is expressed in terms of x_t, and these processes are given, for different initial values, on different probability spaces. As one sees from the construction of the function $b(\varphi, x)$, it also depends on $\bar{x}$. If one can construct the function $b(\varphi, x)$ so that (7) is satisfied for all x, then this dependence vanishes.

We shall now study the question of when the process w(t) in (9) can be chosen to be F_t-measurable. We shall assume that the domain of the function $b(\varphi, x_s)$ in H has for almost all s and x_s one and the same dimension r (we may have $r = +\infty$). We shall denote the closure of the range of $b(\varphi, x_s)$ by H_{x_s}. Let $\hat{H}$ be an r-dimensional Hilbert space and let U_{x_s} be an isometric mapping of H_{x_s} into $\hat{H}$ which is a measurable function of x_s. We set

$$\hat{b}(\varphi, x) = U_x b(\varphi, x).$$

This is a function with values in $\hat{H}$ for which relation (7) is satisfied. Inasmuch as the operator of projection on the closure of the range of $b_s(\varphi) = \hat{b}(\varphi, x_s)$ coincides with the identity operator, then $\eta_t(h) = \hat{\eta}_t(h)$ (we make use here of the notation in Theorem 1), and since $\eta_t(h)$ is F_t-measurable, also $\hat{\eta}_t(t) = (w(t), h)_H$ will also be such, i.e. also the process w(t).

We shall assume that formula (7) is satisfied for all x and that the function $b(\varphi, x)$ is such that the dimension of the range of the mapping $\Phi b(\varphi, x)_{\to}H$ is one and the same for all x. Then, as was shown above, one can assume that it is dense in H. Moreover, let $b(\varphi, x)$ be linear with respect to φ for all x and continuous in Φ in some topology that does not depend on x in which Φ is separable. Then the relation

$$\int_0^t (b(\varphi, x_s), dw_s)_H = \varphi(x_t) - \varphi(x_0) - \int_0^t a(\varphi, x_s)\, ds$$

allows one to define w to be measurable and in a way to be independent of the process x_t. We choose a sequence $\{\varphi_k\}$ that is dense in Φ and let $h_k(x)$ be obtained by orthogonalization (with normalization) of the sequence of vectors $b(\varphi_k, x)$:

$$h_1(x) = b(\varphi_1, x),\ h_2(x) =$$

$$= b(\varphi_2, x) - \frac{b(\varphi_1, x), b(\varphi_2, x))_H}{|b(\varphi_1, x)|^2}\ b(\varphi_1, x),$$

and so forth, whereby we assume the ratio 0/0 to be equal to zero. Some of the $h_k(x)$ may be zero. For all $h \in H$, the equation

$$h = \sum_{k=1}^{\infty} \frac{(h_k(x), h)_H}{|h_k(x)|^2} h_k(x)$$

is valid (we adhere to the same agreement concerning 0/0). Inasmuch as

$$h_k(x)=\sum_{i=1}^{k}\alpha_k(x)b(\varphi_k, x)$$

then

$$\int_0^t (h_k(x_s), dw(s))_H=\sum_{i=1}^{k}\int_0^t \alpha_i(x_s)d\left[\varphi_i(x_s)-\int_0^s a(\varphi_i, x_u)du\right].$$

Hence,

$$(w(t), h)= \\ =\sum_{k=1}^{\infty}\int_0^t \frac{(h_k(x_s), h)_H}{|h_k(x_s)|_H^2}\sum_{i=1}^{k}\alpha_i(x_s)[d\varphi_i(s)-a(\varphi_i x_s)ds]. \tag{10}$$

But Wiener processes, defined by formula (10), are all defined for different initial conditions x_o to be equal on different probability spaces. At the same time, equation (9) can be solved with one and the same Wiener process for different initial conditions. In this connection, the solutions will also be defined on one and the same probability space. We shall show that the Markov process can be realized on one probability space: one can define on one probability space $\{\Omega, \mathfrak{S}, P\}$ a family of random functions $\xi_x(t)$ that measurably depends on a parameter $x \in X$ such that the distribution $\xi_x(t)$ coincides with the distribution of the Markov process x with respect to the measure P_x. In this case, where the process w(t) occurring in equation (9) can be expressed by means of formula (10), $\xi_x(t)$ can be chosen so that

$$d\varphi(\xi_x(t))=a(\varphi, \xi_x(t))dt+(b(\varphi, \xi_x(t)), dw(t))_H, \tag{11}$$

where w(t) is the one and only Wiener process not depending on x.

For the proof of these assertions we shall need the following theorem.

THEOREM 2. Let Y be a complete separable metric space, $(Z, \mathfrak{C})$ a measurable space, $\{\mu_z, z \in Z\}$ a family of probability measures on a σ-algebra of $\mathfrak{B}$-Borel sets from Y, whereby $\mu_z(A)$ is $\mathfrak{C}$-measurable for all $A \in \mathfrak{B}$. Then there exists a function $y(z, \omega)$ defined on $Z \times \Omega$ and measurable relative to $\mathfrak{C} \times \mathfrak{S}$ with values in Y, where $\{\Omega, \mathfrak{S}, P\}$ is some probability space (one can assume that this is the segment [0, 1] with a σ-algebra of Borel sets and Lebesgue measure), such that

$$\mathsf{P}\{y(z, \omega) \in B\} = \mu_z(B) \qquad \forall z \in Z, \quad B \in \mathfrak{B}.$$

There is a proof of this theorem in Gihman and Skorohod's Control of Random Processes (Kiev: Naukova Dumka, 1977, p. 8, Theorem 1.1).

REMARK 1. In the course of the proof of Theorem 1 we used a sequence of measurable random functions $\xi_k(s, \omega)$ defined on the probability space $(\Theta, \mathfrak{C}, \Pi)$, $s \in R_+$, $\omega \in \{\Omega, \mathfrak{F}\}$ such that for all s, ω, ξ_k is a Gaussian sequence with given correlation matrix $\alpha_{kj}(s, \omega)$. In order to establish the existence of such a sequence with the aid of Theorem 2, we take as Y the space of all sequences $\{x^k\}$ with metric $\rho(x, y) = \Sigma 2^{-k}(1 - \exp\{-|x^k - y^k|)$, $\mathfrak{C} - \mathfrak{A}_+ \times \mathfrak{F}$ as $Z - R_+ \times \Omega$, μ_x as the measure defining the Gaussian sequence with correlation matrix $\alpha_{kj}(s, \omega)$.

REMARK 2. Now, let Y be the space of all continuous functions x(t) with values in X with the metric

$$\rho(x(\cdot), y(\cdot)) = \sum_k 2^{-k}(1 - \exp\{\sup_{t \leq k} r_X(x(t), y(t))\}),$$

where r_X is the metric in X, Z = X, μ_z is the measure P_x on Y. Then there exists a function $y(x, \omega)$, measurable with respect to all variables simultaneously, taking on values in Y and such that

$$\mu_x(B) = \mathsf{P}_x(B) = \mathsf{P}\{y(x, \omega) \in B\}$$

for arbitrary Borel set B in Y. Inasmuch as a point of Y is an X-value of the function, then $y(x, \omega) = \xi_x(\cdot, \omega)$, where, for each x, $\xi_x(t, \omega)$ is a continuous process, the function $\xi_x(t, \omega)$ is measurable with respect to all its variables simultaneously. Moreover, for fixed x, to the process $\xi_x(t, \omega)$ there corresponds the measure P_x in the trajectory space.

We show now how to construct the processes $\xi_x(t)$ so that they satsify (11) with one and the same Wiener process w(t). We shall assume that $b(\varphi, x)$ is such that relation (10) is satisfied. We choose some basis $\{h_k\}$ in H and let Z_1 be the set consisting of sequences of numerical continuous functions: $z^{(1)}(t) = \{\alpha_1(t), \alpha_2(t), \ldots\}$ with the topology of uniform coordinatewise convergence on each funite segment. Y is such as in Remark 2. Let μ_x be the measure corresponding to the process $\xi_x(t)$, constructed in Remark 2. For all x, with respect to the measure μ_x there is defined a mapping S from Y into Z_1, defined on the functions x_t by the relation

$$Sx(\cdot)=\{\beta_1(\cdot),\ \beta_2(\cdot),\ \ldots\},\qquad \beta_i(t)=(w(t),\ h_i),$$

where w(t) is defined by equation (10). This mapping is measurable. We set $S\xi_x(\cdot) = \{\beta_i^x(\cdot),\ i = 1, 2, \ldots$. For each x, the sequence $\beta_i^x(t)$ is a sequence of independent Wiener processes with unit dispersion. Inasmuch as Y as well as Z_1 are complete separable metric spaces and the σ-algebras considered in them are Borel, then there exists a regular conditional probability $P\{\xi_x(\cdot) \in B/S\xi_x(\cdot)\}$ on the σ-algebra, i.e. a function $\mu_x,\ z^{(1)}(B)$, $B \in \mathfrak{B}_Y$, $z^{(1)} \in Z_1$, $x \in X$, such that for fixed x, $z^{(1)}$ is a measure on $\mathfrak{B}_Y$ and is measurable relative to $\mathfrak{B} \times \mathfrak{B}_{Z_1}$ ($\mathfrak{B}_{Z_1}$ is a Borel σ-algebra in Z_1), and for which

$$\mu_{x,\,S\xi_x(\cdot)}(B)=\mathsf{P}\,\{\xi_x(\cdot)\in B/S\xi_x(\cdot)\}\tag{12}$$

with probability 1. On the basis of Theorem 2, there exists on some probability space $\{\Omega_1,\ \mathfrak{F}_1,\ P_1\}$ a function $\hat{\xi}(x,\ z^{(1)}(\cdot),\ t)$ such that

$$\mathsf{P}\,\{\hat{\xi}(x,\ z^{(1)}(\cdot),\ \cdot)\in B\}=\mu_{x,\,z^{(1)}(\cdot)}(B).\tag{13}$$

Let $\{\Omega_2,\ \mathfrak{F}_2,\ P_2\}$ be some other probability space on which there is given a sequence of independent Wiener processes $\{\beta_k(t),\ k = 1, 2, \ldots\}$ with unit dispersion. We define on the probability space $\{\Omega_1 \times \Omega_2,\ \mathfrak{F}_1 \times \mathfrak{F}_2,\ P_1 \times P_2\}$ the random function

$$\hat{\xi}(x,\ t)=\hat{\xi}(x,\ \{\beta_1(\cdot),\ \beta_2(\cdot),\ \ldots\},\ t)$$

and a Wiener process w(t) with values in H for which $(w(t),\ h_k) = \beta_k(t)$. Then the joint distribution of $\xi_x(t)$ and $\{\beta_1^x(t),\ \beta_2^x(t),\ \ldots\}$ coincides with the joint distribution of $\hat{\xi}(x,\ t)$ and $\{\beta_1(t),\ \beta_2(t),\ \ldots\}$ inasmuch as the distributions of the sequences $\{\beta_k^x(t),\ k = 1, 2, \ldots\}$ and $\{\beta_k(t),\ k = 1, 2, \ldots\}$ are the same, and the conditional distribution $\hat{\xi}_x(t)$ for fixed sequence $\{\beta_k^x(t),\ k = 1, 2, \ldots\}$ is the same as the conditional distribution $\hat{\xi}(x,\ t)$ for given $\{\beta_k(t),\ k = 1, 2, \ldots\}$ (compare (12) and (13)). From the coincidence of the joint distribution there follows the equation

$$(w(t), h_i) = \sum_{k=1}^{\infty} \int_0^t \frac{(h_k(\xi(x, s)), h_i)}{|h_k(\xi(x, s))|_H^2} \times \\ \times \sum_{j=1}^{k} \alpha_j(\xi(x, s))[d\varphi_j(\xi(x, s)) - a(\varphi_j, \xi(x, s))ds]. \tag{14}$$

Inasmuch as both members of the equation are linear with respect to h_i, then from this it follows that the formula remains valid if h_i is replaced by arbitrary h. Recalling that $h_k(x)$ is expressed linearly in terms of $b(\varphi_i, x)$ and substituting these expressions in (14), we convince ourselves successively of the validity of the equations

$$\int_0^t (b(\varphi_k, \xi(x, s)), dw(s))_H = \\ = (\varphi_k(\xi(x, t)) - \varphi_k(x) - \int_0^t a(\varphi_k, \xi(x, s))ds. \tag{15}$$

Making use of the density of the sequence $\{\varphi_k\}$ in Φ, we can convince ourselves that formula (15) remains valid if φ_k is replaced by any $\varphi \in \Phi$.

We have thus established that under definite assumptions there exists a family of Markov processes $\xi(x, t)$ that satisfy one and the same stochastic differential equation with a Wiener process that does not depend on the initial condition.

3. Existence and Uniqueness of the Solution of a Stochastic Differential Equation. The Smooth Case

Let X and C_0 be, as before, a locally compact space and the space of continuous functions on it that tend to zero at infinity, $D \subset C_0$ is some algebra of functions: this is a linear manifold, closed relative to the multiplication of elements (obviously, C_0 is also an algebra). We assume that two functions are given on $D \times X$: $a(\varphi, x)$ with values in R and $b(\varphi, x)$ with values in some separable (possibly finite-dimensional) Hilbert space. Suppose a locally convex topology can be introduced in D such that it will be separable, and the functions $a(\varphi, x)$ and $b(\varphi, x)$ are continuous in all their variables simultaneously. Moreover, we assume they are linear with respect to φ and satisfy the relation

$$a(\varphi\psi, x) - \varphi(x)a(\psi, x) - \psi(x)a(\varphi, x) = \\ = (b(\varphi, x), b(\psi, x))_H \tag{1}$$

(in the right member, we have written the inner product in H). The meaning of this relation has already been discussed in the preceding section. We shall be interested in the conditions for the existence and uniqueness of the solution of the stochastic differential equation

$$d\varphi(x_t) = a(\varphi, x_t)\,dt + (b(\varphi, x_t), dw(t))_H, \tag{2}$$

where w(t) is a Wiener process in H. Equation (2) is solved with the initial condition $x_o = x$ (one can also consider random initial conditions that do not depend on the process w).

Obviously, in the best case one can define from equation (2) the values of $\varphi(x_t)$ for all $\varphi \in D$. In order to be able to reconstruct the process x_t itself from these values it is necessary (and sufficient) that the functions in D separate points in X, i.e. in order, for an arbitrary pair of points $x_1 \neq x_2$ in X, that it be possible to indicate a function $\varphi \in D$ such that $\varphi(x_1) \neq \varphi(x_2)$. In the sequel, it will always be assumed that D possesses such property. We note that to define x_t it suffices to know $\varphi(x_t)$ for $\varphi \in D_o \subset D$, where D_o is any subset of functions in D, provided that the functions in D_o separate points of X.

We give a simple example of an equation of the form (2) having a solution (and moreover a unique one) for an arbitrary initial value. We shall assume that

$$a(\varphi, x) = A(\varphi, \varphi_1(x), \ldots, \varphi_m(x)),$$
$$b(\varphi, x) = B(\varphi, \varphi_1(x), \ldots, \varphi_m(x)),$$

where $A(\varphi, \xi_1, \ldots, \xi_m)$ and $B(\varphi, \xi_1, \ldots, \xi_m)$ are defined on $D \times R^m$, and that, for each $\varphi \in D$, satisfy with respect to $\xi_1, \ldots, \xi_m$ the Lipschitz condition:

$$\begin{aligned} &|A(\varphi, \xi_1, \ldots, \xi_m) - A(\varphi, \bar{\xi}_1, \ldots, \bar{\xi}_m)| + \\ &\quad + |B(\varphi, \xi_1, \ldots, \xi_m) - B(\varphi, \bar{\xi}_1, \ldots, \bar{\xi}_m)|_H \leqslant \\ &\qquad \leqslant L_\varphi \sum_{k=1}^{m} |\xi_k - \bar{\xi}_k|. \end{aligned}$$

In this case, for $\varphi_1(x_t), \ldots, \varphi_n(x_t)$ we obtain a system of stochastic equations:

$$\begin{aligned} d\varphi_k(x_t) = {} & A(\varphi_k, \varphi_1(x_t), \ldots, \varphi_m(x_t))\,dt + \\ & + (B(\varphi_k, \varphi_1(x_t), \ldots, \varphi_m(x_t)), \quad dw(t))_H, \end{aligned} \tag{3}$$

to which the classical theory is applicable, from which it follows that there exists a unique collection of functions $\eta_1(t), \ldots, \eta_m(t)$ satisfying relation (3) if we substitute here $\eta_k(t)$ in place of $\varphi_k(x_t)$, whereby they will be measureable relative to the filtration of σ-algebras generated by the Wiener process w(t); for the remaining $\varphi \in D$, $\varphi(x_t)$

is defined from the equations

$$\varphi(x_t) = \varphi(x) + \int_0^t A(\varphi, \eta_1(s), \ldots, \eta_m(s))\, ds + \\ + \int_0^t (B(\varphi, \eta_1(s), \ldots, \eta_m(s)), dw(s))_H$$

(in the right member there occur only known quantities).

In the proof of the existence and uniqueness of the solution of equation (2) it is convenient to make use of solutions on random intervals of the form $[0, \tau)$, where τ is the moment of emergence from some neighbourhood, containing the initial point and having compact closure. We shall assume that for each point x there exists a neighbourhood U_x with compact closure such that the solution of (2) is unique up to the moment of emergence from U_x. This means that if $x(t)$ and $\bar{x}(t)$ are two solutions of (2), satisfying the initial condition $x(0) = \bar{x}(0) = x$, and $\tau(\bar{\tau})$ are the moments of exit of the process $x(t)(\bar{x}(t))$ from the neighborhood U_x, then $x(t) = \bar{x}(t)$ for $t \leqq \tau \wedge \bar{\tau}$. (Only continuous solutions are considered.) In particular, this condition, as is well known, is satisfied if there does not exist a solution with initial condition x. We shall say that in this case solution (2) is locally unique. We shall show that local uniqueness implies uniqueness. In fact, let $x(t)$ and $\bar{x}(t)$ be two solutions satisfying the initial condition $\bar{x}(0) = x(0) = x$. We denote by F_t the flow of σ-algebras generated by a Wiener process and the solutions $x(t)$ and $\bar{x}(t)$. Let $\tau = \inf[s: x(s) \neq \bar{x}(s)]$. This is the stoppage time relative to the filtration F_t, to which the process $w(t)$ is adapted. Therefore, the process $w(t + \tau) - w(\tau)$ does not depend on F_τ and is a Wiener process. Making use of the relation

$$\varphi(x(t+\tau)) - \varphi(x(\tau)) = \\ = \int_0^t a(\varphi, x(\tau+s))\, ds + \int_0^t (b(\varphi, x(\tau+s)), \\ d[w(s+\tau) - w(\tau)])$$

and the analogous relation for $\bar{x}(\cdot)$, one can assert that $x(t + \tau) = \bar{x}(t + \tau)$ for $t < \zeta$, where ζ is the exit time of the process $x(t + \tau)$ from the neighborhood $U_{x(\tau)}$. It follows from the continuity of the process $x(\cdot)$ that $\zeta > 0$. But, by the definition of τ, there exist arbitrarily small t such that $x(\tau + t) \neq \bar{x}(\tau + t)$. Thus, the assumption that τ is finite leads to a contradiction and hence $x(s) = \bar{x}(s)$ for all $s \geqq 0$.

If, for each $\bar{x}$ there exists a neighborhood $U_{\bar{x}}$ such that one can find coefficients $\bar{a}(\varphi, x)$ and $\bar{b}(\varphi, x)$ such that equation (2) has a solution with these coefficients (i.e. for $a = \bar{a}$ and $b = \bar{b}$) and initial condition $\bar{x}$, and, moreover, $\bar{a}(\varphi, x) = a(\varphi, x)$, $\bar{b}(\varphi, x) = b(\varphi, x)$ for $x \in U_x$, then we shall say that the equation has a local solution.

In this case, one can construct successively a solution up to the moment of exit $x(t)$ from all compacta, i.e. a moment ζ such that $x(t) \to \infty$ as $t \to \zeta$. If it turns out that $\zeta = +\infty$, then the existence of the solution will follow from the existence of the local solution. We give one sufficient condition in order that $\zeta = +\infty$.

LEMMA 1. Suppose that for equation (2) there exists a local solution and that one can find a function Φ satisfying the conditions: (a) $\Phi(x) \geq 0$, $\Phi(x) \to \infty$ as $x \to \infty$; (b) for every compactum $K \in X$ there exists a function $\varphi \in D$ for which $\Phi = \varphi$ for $x \in K$ and $a(\varphi, x) \leq \alpha\varphi + \beta$, where the constants $\alpha > 0$ and $\beta > 0$ do not depend on the compactum K. Then equation (2) has a solution.

Proof. It will be sufficient to prove that, for all $t > 0$, $P\{\zeta > t\} = 1$. Let τ be the moment of exit from the compactum K. Inasmuch as

$$d(e^{-\alpha t}\varphi(x_t)) = e^{-\alpha t}[-\alpha\varphi(x_t) + a(\varphi, x_t)]\,dt + $$
$$+ e^{-\alpha t}(b(\varphi, x_t), d(w(t)) \leqslant e^{-\alpha t}\beta\,dt + e^{-\alpha t}(b(\varphi, x_t), dw(t)),$$
$$d\left(e^{-\alpha t}\left[\varphi(x_t) + \frac{\beta}{\alpha}\right]\right) \leqslant e^{-\alpha t}(b(\varphi, x_t), dw(t)),$$

then, for $t_1 < t_2 \leq \tau$,

$$e^{-\alpha t_2}\left(\Phi(x_{t_2}) + \frac{\beta}{\alpha}\right) - e^{-\alpha t_1}\left(\Phi(x_{t_1}) + \frac{\beta}{\alpha}\right) \leqslant$$
$$\leqslant \int_{t_1}^{t_2} e^{-\alpha s}(b(\varphi, x_s), dw(s)).$$

We denote

$$e^{-\alpha t}\left(\Phi(x_t) + \frac{\beta}{\alpha}\right) = \eta_t.$$

Then $\eta_{t\wedge\tau}$ is a supermartingale:

$$\mathsf{M}(\eta_{t_2\wedge\tau}/\mathcal{F}_{t_1\wedge\tau}) \leqslant$$
$$\leqslant \eta_{t_1\wedge\tau} + \mathsf{M}\left(\int_{t_1\wedge\tau}^{t_2\wedge\tau} e^{-\alpha s}(b(\varphi, x_s), dw(s))/\mathcal{F}_{t_1\wedge\tau}\right) = \eta_{t_1\wedge\tau}.$$

We choose a sequence of compacta K_n so that $\bigcup K_n = X$ and $K_n \subset K_{n+1}$. If τ_n is the moment of exit from the compactum K_n, then $\tau_n \uparrow \zeta$, and hence η_t is a supermartingale on $[0, \zeta)$. Therefore, the sequence

$\eta_{t\wedge\tau_n}$ will be a supermatringale for arbitrary t.

Inasmuch as $|M_{\eta_{t\wedge\tau_n}} \leq M\eta_{t\wedge\tau_1} \leq M\eta_0 = \Phi(\bar{x}) + \beta/\alpha$. ($\bar{x}$ is the initial value of x_t), then

$$M\Phi(x_{t\wedge\tau_n}) \leqslant e^{\alpha t}\left[\Phi(x) + \frac{\beta}{\alpha}\right] - \frac{\beta}{\alpha}.$$

Let $K_n = \{x : \Phi(x) \leq n\}$. Then $\Phi(x_{\tau_n}) = n$ and consequently

$$nP\{\tau_n \leqslant t\} \leqslant e^{\alpha t}\left[\Phi(x) + \frac{\beta}{\alpha}\right],$$
$$P\{\zeta \leqslant t\} \leqslant \frac{1}{n} e^{\alpha t}\left[\Phi(x) + \frac{\beta}{\alpha}\right],$$

for arbitrary n. This completes the proof of the lemma.

The connection just considered between the local existence and the local uniqueness, on the one hand, and the existence and uniqueness of the solution of equation (2) on the other hand, allows one to reduce the study of equation (2) to the study the same kind of equation but for a compact phase space. Let K be an arbitrary compact set in X and let X_1 be a compact set such that all points of K are interior points of X_1. We shall define the functions $\tilde{a}(\varphi, x)$ and $\tilde{b}(\varphi, x)$, $\varphi \in \tilde{D}$, so that for all $\varphi \in D$ on X_1 there exists a $\tilde{\varphi} \in \tilde{D}$ such that $\varphi = \tilde{\varphi}$ for $x \in K$ and $\tilde{a}(\tilde{\varphi}, x) = a(\varphi, x)$, $\tilde{b}(\tilde{\varphi}, x) = b(\varphi, x)$ for $x \in K$. If $\tilde{x}_t$ is the solution of the equation

$$d\tilde{x}_t = \tilde{a}(\tilde{\varphi}, \tilde{x}_t)\,dt + (\tilde{b}(\tilde{\varphi}, \tilde{x}_t), dw(t))_H, \quad \tilde{\varphi} \in \tilde{D},$$

and τ is the moment of exit from K, then, for $t \leq \tau$, $\tilde{x}_t$ will also be a solution of equation (2). In the sequal of this section it is assumed that X is compact.

UNIQUENESS THEOREM. Suppose there exists a set $D_0 \subset D$ of functions, separating points in X, such that for some l the following condition is satisfied: for arbitrary two random quantities ξ and η with values in X, with $\varphi \in D_0$,

$$M\left(|a(\varphi, \xi) - a(\varphi, \eta)|^2 + |b(\varphi, \xi) - b(\varphi, \eta)|_H^2\right) \leqslant$$
$$\leqslant l \sup_{\psi \in D_0} M|\psi(\xi) - \psi(\eta)|^2.$$

Then the solution of equation (2) is unique.

<u>Proof</u>. Suppose x_t and $\bar{x}_t$ are two solutions of equation (2) with the same initial value. Then, for $\varphi \in D_0$

$$\mathsf{M}(\varphi(x_t)-\varphi(\tilde{x}_t))^2=\mathsf{M}\left(\int_0^t[a(\varphi,\ x_s)-a(\varphi,\ \tilde{x}_s)]\,ds+\right.$$

$$\left.+\int_0^t(b(\varphi,\ x_s)-b(\varphi,\ \tilde{x}_s),\ dw(s)\right)^2\leqslant$$

$$\leqslant 2t\int_0^t\mathsf{M}\,|a(\varphi,\ x_s)-a(\varphi,\ \tilde{x}_s)|^2\,ds+$$

$$+\int_0^t\mathsf{M}\,|b(\varphi,\ x_s)-b(\varphi,\ \tilde{x}_s)\,|_H^2\,ds\leqslant$$

$$\leqslant(2t+1)\,l\int_0^t\sup_{\psi\in D_0}\mathsf{M}\,|\,\psi(x_s)-\psi(\tilde{x}_s)\,|^2\,ds.$$

Consequently, for $t \leqq T$,

$$\sup_{\varphi\in D_0}\mathsf{M}\,|\,\varphi(x_t)-\varphi(\tilde{x}_t)\,|^2\leqslant$$

$$\leqslant(2T+1)\,l\int_0^t\sup_{\varphi\in D_0}\mathsf{M}\,|\,\varphi(x_s)-\varphi(\tilde{x}_s)\,|^2\,ds.$$

Setting

$$\alpha(t)=\sup_{\varphi\in D_0}\mathsf{M}\,|\,\varphi(x_t)-\varphi(\tilde{x}_t)\,|^2,\ l_1=(2T+1)\,l,$$

we shall have

$$\alpha(t)\leqslant l_1\int_0^t\alpha(s)\,ds,$$

whence $\alpha(t) = 0$. This completes the proof of the theorem.

REMARK 1. Condition (4) is a generalization of the Lipschitz condition on the coefficients of the equation if the space is Euclidean. In this case, H is also finite-dimensional,

$$a(\varphi,\ x)=\sum\alpha_i(x)\frac{\partial\varphi}{\partial x_i}+\sum\beta_{ij}(x)\frac{\partial^2\varphi}{\partial x_i\partial x_j};$$

$$b(\varphi,\ x)=\sum\frac{\partial\varphi}{\partial x_i}b_i(x),$$

where $b_i(x) \in H$. In the role of D_0, we take the coordinate functions $\varphi_i(x) = x_i$. Then

$$|a(\varphi_i, x)-a(\varphi_i, y)|^2+|b(\varphi_i, x)-b(\varphi_i, y)|_H^2=$$
$$=|\alpha_i(x)-\alpha_i(y)|^2+|b_i(x)-b_i(y)|_H^2.$$

If the right side does not surpass $l|x-y|^2$ for all x, y, then condition (4) will be satisfied.

REMAR 2. As we discussed above, it suffices to prove local uniqueness. Therefore, it suffices to require that for each closed neighborhood one can find a set D_0 such that (4) is satisfied for the random variables ξ and η, taking on values from this neighborhood.

We now go over to the proof of existence. We need some auxiliary propositions concerning analysis.

LEMMA 2. Let X be a compactum with distance r, let $\{\varphi_n\}$ be a sequence of functions from C_X, separating points of X, $D \subset C_X$ some compact set. Then, for each $\varepsilon > 0$, one can find an n such that for arbitrary $\varphi \in D$ there exists a continuous function $F(s_1, \ldots, s_n)$ on R^n for which

$$\sup_{x\in X}|\varphi(x)-F(\varphi_1(x), \ldots, \varphi_n(x))|\leqslant\varepsilon.$$

Proof. We denote by $K_n \subset R^n$ the image of X under the mapping $x \to (\varphi_1(x), \ldots, \varphi_n(x))$. Obviously, K is a compactum and K_n is the image of K_{n+1} under projection on R^{n+1}: $(s_1, \ldots, s_{n+1}) \to (s_1, \ldots, s_n)$. We define the sets

$$\Delta_{s_1,\ldots,s_n}=\{x: \varphi_1(x)=s_1, \ldots, \varphi_n(x)=s_n\}.$$

Let $d(\Delta_{s_1, \ldots, s_n})$ be the diameter of the set $\Delta_{s_1, \ldots, s_n}$; $\delta_n(x) = d(\Delta_{\varphi_1(x), \ldots, \varphi_n(x)})$. For every sequence of points $\{s_k\}$, $\bigcap_n \Delta_{s_1, \ldots, s_n}$ is either empty (if for some n the point $(s_1, \ldots, s_n)$ does not belong to K_n), or it consists of one point x (if $s_k = \varphi_k(x)$, such a point x is unique by virtue of the fact that the $\{\varphi_n\}$ separate points of X). Since $\Delta_{s_1, \ldots, s_n} \supset \Delta_{s_1, \ldots, s_{n+1}}$, then $d(\Delta_{s_1, \ldots, s_n}) \to 0$ as $n \to \infty$ (if this were not so, then the set $\bigcap_n \Delta_{s_1, \ldots, s_n}$ would contain at least two points). Hence $\delta_n(x) \downarrow 0$. We shall show that the function $\delta_n(x)$ is upper semicontinuous, i.e. that

$$\overline{\lim_{x\to x_0}}\,\delta_n(x)\leqslant\delta_n(x_0).$$

Let

$$\Delta^{\gamma}_{s_1,\dots,s_n} = \{x\colon |\varphi_1(x) - s_1| \leqslant \gamma, \ \dots, |\varphi_n(x) - s_n| \leqslant \gamma\}.$$

Then

$$\Delta_{s_1,\dots,s_n} = \bigcap_{\gamma>0} \Delta^{\gamma}_{s_1,\dots,s_n}$$

and

$$d(\Delta_{s_1,\dots,s_n}) = \lim_{\gamma \downarrow 0} d(\Delta^{\gamma}_{s_1,\dots,s_n}).$$

Therefore, for each $\varepsilon > 0$ one can find a $\gamma > 0$ such that

$$d(\Delta^{\gamma}_{\varphi_1(x_0),\dots,\varphi_n(x_0)}) \leqslant d(\Delta_{\varphi_1(x_0),\dots,\varphi_n(x_0)}) + \varepsilon.$$

On the other hand, there exists a $\delta > 0$ such that for $r(x, x_o) < \delta$, $|\varphi_k(x) - \varphi_k(x_o)| < \gamma$, $k = 1, \dots, n$. But then

$$\Delta_{\varphi_1(x),\dots,\varphi_n(x)} \subset \Delta^{\gamma}_{\varphi_1(x_0),\dots,\varphi_n(x_0)},$$
$$\delta_n(x) \leqslant d(\Delta^{\gamma}_{\varphi_1(x_0),\dots,\varphi_n(x_0)}) \leqslant \delta_n(x_0) + \varepsilon.$$

And the upper semicontinuity of $\delta_n(x)$ follows from this inequality.

By Dini's theorem, a sequence of upper semicontinuous functions on a compactum that decreases monotonously to zero converges to zero uniformly. Hence, for each $\rho > 0$ one can find an n such that $\sup_{x\in X} \delta_n(x) < \rho$.

We choose an arbitrary point $\hat{x}_{s_1, \dots, s_n}$ in $\Delta_{s_1, \dots, s_n}$ and set

$$F_1(s_1, \dots, s_n) = \varphi(\hat{x}_{s_1,\dots,s_n}).$$

Then

$$|F_1(\varphi_1(x), \dots, \varphi_n(x)) - \varphi(x)| \leqslant$$
$$\leqslant \sup_{y\in\Delta_{\varphi_1(x),\dots,\varphi_n(x)}} |\varphi(y) - \varphi(\hat{x}_{\varphi_1(x),\dots,\varphi_n(x)})| \leqslant$$
$$\leqslant \sup_{r(x_1, x_2)<\rho} |\varphi(x_1) - \varphi(x_2)|.$$

We estimate the oscillation of the function $F_1(s_1, \dots, s_n)$ at the point $(s_1^o, \dots, s_n^o) \in K_n$. We take $\gamma > 0$ and suppose the point $(s_1, \dots, s_n) \in K_n$ lies at a distance not surpassing γ from the point $(s_1^o, \dots, s_n^o)$.

Then

$$\Delta_{s_1, \ldots, s_n} \subset \Delta^{\gamma}_{s_1^0, \ldots, s_n^0}.$$

Inasmuch as by the choice of γ one can make the diameter of $\Delta_{s_1^0, \ldots, s_n^0}$, arbitrarily close to the diameter of $\Delta_{s_1^0, \ldots, s_n^0}$, then, for sufficiently small γ, $d(\Delta^{\gamma}_{s_1^0, \ldots, s_n^0}) < \rho$. Hence, $r(\hat{x}_{s_1, \ldots, s_n}, \hat{x}_{s_1^0, \ldots, s_n^0}) < \rho$ both of these points belong to $\Delta^{\gamma}_{s_1^0, \ldots, s_n^0}$) and

$$\begin{aligned} |F_1(s_1, \ldots, s_n) - F_1(s_1^0, \ldots, s_n^0)| &= \\ = |\varphi(\hat{x}_{s_1, \ldots, s_n}) - \varphi(\hat{x}_{s_1^0, \ldots, s_n^0})| &\leqslant \\ \leqslant \sup_{r(x_1, x_2) < \rho} |\varphi(x_1) - \varphi(x_2)|. \end{aligned}$$

Thus, the oscillation of the function $F(s_1, \ldots, s_n)$ at an arbitrary point does not surpass the quantity

$$\omega_\rho(\varphi) = \sup_{r(x_1, x_2) < \rho} |\varphi(x_1) - \varphi(x_2)|.$$

Therefore, for each $\varepsilon_1 > 0$ one can find a continuous function $F(s_1, \ldots, s_n)$, defined on K_n, such that

$$\sup_{(s_1, \ldots, s_n) \in K_n} |F(s_1, \ldots, s_n) - F_1(s_1, \ldots, s_n)| < 2(\omega_\rho(\varphi) + \varepsilon_1).$$

This function can be constructed, for example, in the following way. Inasmuch as the oscillation of the function at a point is the greatest lower bound of the oscillation of the function in a neighborhood containing the point, and K_n is a compactum, one can find a finite number of spheres in R^n covering K_n such that the oscillation of the function φ in each sphere does not surpass $\omega_\rho(\varphi) + \varepsilon_1$. We now choose a finite set of points so that each sphere has at least one point and in addition a covering and the set of points should contain a minimal number of elements. A polygonal function whose values at the point $(\bar{s}_1, \ldots, \bar{s}_n)$ of the selected set coincide with the values of F_1 at one of the points of the sphere to which $(\bar{s}_1, \ldots, \bar{s}_n)$ belongs will be the one sought.

Since D is an equicontinuous set of functions, then choosing $\rho > 0$

so that $\omega_\rho(\varphi) < 3\varepsilon/4$ for $\varphi \in D$, and then n so that $\sup_{x \in X} \delta_n(x) < \rho$, we shall have for $\varepsilon_1 = \varepsilon/8$:

$$|\varphi(x) - F(\varphi_1(x), \ldots, \varphi_n(x))| < \\ < |\varphi(x) - F_1(\varphi_1(x), \ldots, \varphi_n(x))| + 2\omega_\rho'(\varphi) + 2\varepsilon_1 \leqslant \\ \leqslant 3\omega_\rho(\varphi) + \varepsilon/4 \leqslant \varepsilon.$$

This completes the proof of the lemma.

REMARK 1. Without any modification of the proof, the lemma carries over to the case when D consists of functions taking on values in the Hilbert space H. One must only understand $\omega_\rho(\varphi)$ to be the quantity

$$\omega_\rho(\varphi) = \sup_{r(x, y) < \rho} |\varphi(x) - \varphi(y)|_H.$$

REMARK 2. A function from K_n into X, $(s_1, \ldots, s_n) \to \hat{x}_{s_1, \ldots, s_n}$, can be constructed with the aid of the axiom of choice. It represents a 'branch' of the function inverse to the function

$$x \to (\varphi_1(x), \ldots, \varphi_n(x)),$$

inasmuch as $\varphi_k(\hat{x}_{s_1, \ldots, s_n}) = s_k$, $k = 1, \ldots, s_n$. No mention is made of any regularity conditions of such a function. In the following lemma it will be shown that this function can be chosen to be a Borel function.

LEMMA 3. Suppose $K_n \subset R^n$ is the image of the copactum X under the mapping $x \to (\varphi_1(x), \ldots, \varphi_n(x))$. Then there exists a Borel function $\hat{x}_n(s_1, \ldots, s_n)$, defined on the compactum K_n and with values in X, for which $\varphi(\hat{x}_n(s_1, \ldots, s_n)) = s_k$ for $k = 1, \ldots, n$, $(s_1, \ldots, s_n) \in K_n$.

<u>Proof</u>. We consider the following function, defined on the product of compacta $X \times K_n$:

$$G(x; s_1, \ldots, s_n) = \sum_1^n (\varphi_k(x) - s_k)^2.$$

This function is continuous and $\inf_x G(x; s_1, \ldots, s_n) = 0$ for all $(s_1, \ldots, s_n) \subset K_n$. Therefore, there exists a Borel function $\hat{x}_n(s_1, \ldots, s_n)$ for which $G(\hat{x}_n(s_1, \ldots, s_n); s_1, \ldots, s_n) = 0$ (see Lemma 1.4 in Gihman and Skorohod's <u>Control Random Processes</u> (Kiev: Naukova Dumka, 1977 (Russian)*). This completes the proof of the lemma.

* English translation: Springer

REMARK. As follows from the proof of Lemma 2, the oscillation of the function $\hat{x}_n(s_1, \ldots, s_n)$ at any point of K_n does not surpass

$$\sup_{(s_1, \ldots, s_n)} d(\Delta_{s_1, \ldots, s_n}) = \sup_x \delta_n(x)$$

and this quantity tends to zero as $n \to \infty$.

EXISTENCE THEOREM. Suppose there exist a countable set $D_o \subset D$, $D_o = \{\varphi_n, n = 1, 2, \ldots\}$, separating points of X, and a sequence $\varepsilon_n \downarrow 0$ such that for some l the following condition is satisfied: for two arbitrary and random variables ξ and η with values in X,

$$\mathsf{M}\left(|a(\varphi_k, \xi) - a(\varphi_k, \eta)|^2 + |b(\varphi_k, \xi) - b(\varphi_k, \eta)|_H^2\right) \leqslant \leqslant l \sup_{j \leqslant n} \mathsf{M}\,|\varphi_j(\xi) - \varphi_j(\eta)|^2 + \varepsilon_n$$

for $k \leq n$. Then, for all $x \in K$, there exists a solution of equation (2) with initial condition $x_o = x$.

Proof. It follows from Lemma 2 and Remark 1 that there exist sequences of functions $\{F_{n,k}(s_1, \ldots, s_n)\}$, $\{B_{n,k}(s_1, \ldots, s_n)\}$, defined on the compactum $K_n \subset R^n$ (as pointed out in Lemma 2) and taking on values in R and H respectively, for which

$$\sup_{x \in X}\{|F_{n,k}(\varphi_1(x), \ldots, \varphi_n(x)) - a(\varphi_k, x)| + + |B_{n,k}(\varphi_1(x), \ldots, \varphi_n(x)) - b(\varphi_k, x)|_H\} \to 0.$$

Since continuous functions on compacta can be uniformly approximated by differentiable functions with bounded derivatives, then, without loss of generality, one can assume that the functions $F_{n,k}(s_1, \ldots, s_n)$ and $B_{n,k}(s_1, \ldots, s_n)$ satisfy the Lipschitz condition with respect to s (the constant in the Lipschitz condition can depend on n and k). We consider the following system of stochastic equations in R^n:

$$d\sigma_k^{(n)}(t, \omega) = F_{n,k}(\sigma_1^{(n)}(t, \omega), \ldots, \sigma_n^{(n)}(t, \omega))\, dt + + (B_{n,k}(\sigma_1^{(n)}(t, \omega), \ldots, \sigma_n^{(n)}(t, \omega)), dw(t))_H. \tag{5}$$

Inasmuch as the coefficients of the system (5) satisfy the Lipschitz condition, this system has moreover a unique solution. Furthermore, let $\hat{x}_n(s_1, \ldots, s_n)$ be a Borel function for which

$$\varphi_k(\hat{x}_n(s_1, \ldots, s_n)) = s_k, \; k = 1, \ldots, n; \quad (s_1, \ldots, s_n) \in K_n,$$

the existence of such a function was established in Lemma 3. We shall denote by $x(t, \omega)$ a random process with values in X defined by the equation

$$x_n(t, \omega) = \hat{x}_n(\sigma_1^{(n)}(t, \omega), \ldots, \sigma_n^{(n)}(t, \omega)).$$

Furthermore, let $a_n(\varphi_k, x) = F_{n,k}(\varphi_1(x), \ldots, \varphi_n(x))$, $b_n(\varphi, x) = B_{n,k}(\varphi_1(x), \ldots, \varphi_n(x))$ for $k \leq n$. We denote $D_n = \{\varphi_1, \ldots, \varphi_n\}$. From (5) there follows such a relation for the process $x_n(t, \omega)$:

$$\begin{gathered} d\varphi(x_n(t, \omega)) = a_n(\varphi, x_n(t, \omega))\,dt + \\ + (b_n(\varphi, x_n(t, \omega)), dw(t))_H, \quad \varphi \in D_n, \\ x_n(0, \omega) = \hat{x}_n(\varphi_1(x), \ldots, \varphi_n(x)). \end{gathered} \tag{6}$$

We denote $\delta_n = \sup_{y \in x} \delta_n(y)$. As was shown in the proof of Lemma 2, $\delta_n \downarrow 0$, and it follows from Lemma 3 that

$$r(\hat{x}_n(\varphi_1(x), \ldots, \varphi_n(x)), x) \leqslant \delta_n.$$

For $k \leq n$ and $m > n$, we estimate the difference $\varphi_k(x_n(t, \omega)) - \varphi_k(x_m(t, \omega))$. We have

$$\begin{gathered} \varphi_k(x_n(t, \omega)) - \varphi_k(x_m(t, \omega)) = \varphi_k(x_n(0, \omega)) - \varphi_k(x_m(0, \omega)) + \\ + \int_0^t [a_n(\varphi_k, x_n(s, \omega)) - a_m(\varphi_k, x_m(s, \omega))]\,ds + \\ + \int_0^t (b_n(\varphi_k, x_n(s, \omega)) - b_m(\varphi_k, x_m(s, \omega)), dw(s))_H. \end{gathered}$$

In estimating the integrals we shall make use of the notation

$$\sup_x \{|a_n(\varphi_k, x) - a(\varphi_k, x)|^2 + |b_n(\varphi_k, x) - b(\varphi_k, x)|_H^2\} = \varepsilon_{n,k}.$$

By construction, $\varepsilon_{nk} \to 0$ as $n \to \infty$ for all k. We have

$$\begin{aligned} \int^t |a_n(\varphi_k, x_n(s, \omega)) - a_m(\varphi_k, x_m(s, \omega))|\,ds = \\ = \int_0^t |a(\varphi_k, x_n(s, \omega)) - a(\varphi_k, x_m(s, \omega))|\,ds + \\ + \int_0^t [a_n(\varphi_k, x_n(s, \omega)) - a(\varphi_k, x_n(s, \omega))]\,ds + \\ + \int_0^t [a(\varphi_k, x_m(s, \omega)) - a_m(\varphi_k, x_m(s, \omega))]\,ds = \\ = \int_0^t [a(\varphi_k, x_n(s, \omega)) - a(\varphi_k, x_m(s, \omega))|\,ds + O\left(t\left(\sqrt{\varepsilon_{nk}} + \sqrt{\varepsilon_{mk}}\right)\right). \end{aligned}$$

Analogously, we can obtain the estimate

$$\int_0^t (b_n(\varphi_k, x_n(s,\omega)) - b_m(\varphi_k, x_m(s,\omega)), dw(s))_H =$$
$$= \int_0^t (b(\varphi_k, x_n(s,\omega)) - b(\varphi_k, x_m(s,\omega)), dw(s))_H + \alpha,$$

where $M\alpha^2 = O(t(\varepsilon_{nk} + \varepsilon_{mk}))$. Finally, noting that $\varphi_k(x_n(0, \omega)) = \varphi_k(x) = \varphi_k(x_m(0, \omega))$, we convince ourselves that, for arbitrary $T > 0$, there exists a c_T such that, for $t \leq T$, the estimate

$$M[\varphi_k(x_n(t,\omega)) - \varphi_k(x_m(t,\omega))]^2 \leqslant c_T(\varepsilon_{nk} + \varepsilon_{mk} +$$
$$+ M\int_0^t (|a(\varphi_k, x_n(s,\omega)) - a(\varphi_k, x_m(s,\omega))|^2 +$$
$$+ |b(\varphi_k, x_n(s,\omega)) - b(\varphi_k, x_n(s,\omega)|_H^2)\, ds.$$

is valid. Consequently, for arbitrary $r < n < m$

$$\sup_{k\leqslant r} M[\varphi_k(x_n(t,\omega)) - \varphi_k(x_m(t,\omega))]^2 \leqslant c_T(\sup_{k\leqslant r}(\varepsilon_{nk} + \varepsilon_{mk}) +$$
$$+ l\int_0^t [\sup_{j\leqslant r} M(\varphi_j(x_n(s,\omega)) - \varphi_j(x_m(s,\omega)))^2 + \varepsilon_r]\, ds).$$

It follows from the preceding inequality that for some c_1 and c_2, depending on T,

$$\sup_{k\leqslant r} M[\varphi_k(x_n(t,\omega)) - \varphi_k(x_m(t,\omega))]^2 \leqslant$$
$$\leqslant c_1\left(\sup_{k\leqslant r}(\varepsilon_{nk} + \varepsilon_{mk}) + \varepsilon_r\right) e^{c_2 t}.$$

Therefore, for arbitrary r,

$$\lim_{n,m\to\infty} \sup_{k\leqslant r} M[\varphi_k(x_n(t,\omega)) - \varphi_k(x_m(t,\omega))]^2 = O(\varepsilon_r).$$

Passaging to the limit as $r \to \infty$, we convince ourselves that, for all k and T,

$$\lim_{n,m\to\infty} \sup_{t\leqslant T} M[\varphi_k(x_n(t,\omega)) - \varphi_k(x_m(t,\omega))]^2 = 0. \tag{7}$$

It follows from this relation that $x_n(t, \omega)$ as $n \to \infty$ converges in X in probability to some limit. In fact, we introduce a new metric in X:

$$r_1(x, y) = \sum \alpha_k |\varphi_k(x) - \varphi_k(y)|,$$

where $\alpha_k > 0$ is chosen so that $\Sigma\alpha_k||\varphi_k|| < \infty$ (if $r_1(x, y) = 0$, then $x = y$; this follows from the fact that $\{\varphi_k\}$ separates points in X). Obviously, $r_1(x, y) \to 0$ as $r(x, y) \to 0$. The converse is also valid: if $r_1(x, y) \to 0$ then also $r(x, y) \to 0$ (if this were not the case, then by virtue of the compactness of X one could find sequences $x_n \to x_o$, $y_n \to y_o$ such that $r(x_n, y_n)$ would not tend to zero, i.e. $(r(x_o, y_o) > 0$, and $r_1(x_n, y_n) \to 0$, i.e. $\varphi_k(x_o) = \varphi_k(y_o)$ for all k). It follows from (6) that

$$\lim_{n, m\to\infty} \sup_{t\leq T} \mathsf{M} r_1(x_n(t, \omega), x_m(t, \omega)) = 0,$$

i.e. $x_n(t, \omega)$ converges in probability with respect to the metric r_1. But then $x_n(t, \omega)$ also converges in probability with respect to the metric r. We denote this limit by $x(t, \omega)$. Since $a_n(\varphi_k, x)$ converges uniformly to $a(\varphi_k, x)$ and $x_n(t, \omega)$ converges in probability to $x(t, \omega)$ and $a(\varphi_k, x)$ is a continuous function, then $a_n(\varphi_k, x_n(t, \omega)) \to a(\varphi_k, x(t, \omega))$ in probability. In exactly the same way, $b_n(\varphi_k, x_n(t, \omega)) \to b(\varphi_k, x(t, \omega))$ in probability in H. Rewriting relation (6) in an integrated form,

$$\varphi_k(x_n(t, \omega)) = \varphi_k(x) + \int_0^t a_n(\varphi_k, x_n(s, \omega))\,ds + \int_0^t (b_n(\varphi_k, x_n(s, \omega)), dw(s))_H, \quad k \leqslant n,$$

and then passing to the limit as $n \to \infty$, we convince ourselves that

$$\varphi_k(x(t, \omega)) = \varphi_k(x) + \int_0^t a(\varphi_k, x(s, \omega))\,ds + \int_0^t (b(\varphi_k, x(s, \omega)), dw(s))_H$$

for all k. If $x(t, \omega)$ is chosen to be separable, then $\varphi_k(x(t, \omega))$ is continuous for all k. Consequently, $x(t, \omega)$ is continuous with respect to the metric r_1 and it is therefore also continuous with respect to the metric r. This completes the proof of the theorem.

REMARK. In this case when the conditions of the existence theorem as well as of the uniqueness theorem are satisfied, then the solution will be a homogeneous Markov process (this is established in the same way as in finite-dimensional Euclidean space). Inasmuch as for $\varphi \in D$,

$$M\varphi(x(t, \omega)) = \varphi(x(0, \omega)) + \int_0^t Ma(\varphi, x(s, \omega))\, ds,$$

then these functions occur in the domain of the infinitesimal operator of the process and a (φ, x) is the value of this operator. It is easy to convince oneself that, for arbitrary $\varphi_1, \ldots, \varphi_n \in D$ and a twice continuously differentiable function on R^n, $F(s_1, \ldots, s_n)$ is the infinitesimal operator of the process defined on the function $F(\varphi_1(x), \ldots, \varphi_n(x))$ and coincides with the expression

$$\sum_{k=1}^{n} \frac{\partial}{\partial \varphi_k} F \cdot a(\varphi_k, x) + \frac{1}{2} \sum_{k,j=1}^{n} \frac{\partial^2}{\partial \varphi_k \partial \varphi_j} F \cdot (b(\varphi_k, x), b(\varphi_j, x))_H.$$

We shall analyze the connection between the conditions of the existence theorem and of the uniqueness theorem. Inasmuch as the space C_X of continuous functions on the compactum X is separable, then from each set of continuous functions one can choose a countable everywhere dense set. In addition, if one replaces the set D_o by some set $\hat{D}_o$ dense in it, then condition (4) of the uniqueness theorem will be satisfied. Thus, D_o can also be assumed countable in the uniqueness theorem. We assume that $D_o = \{\varphi_n, n = 1, 2, \ldots\}$. If the conditions of the existence theorem are satisfied, then the conditions of the uniqueness theorem will also be satisfied since, for arbitrary k,

$$\begin{aligned} M\{|a(\varphi_k, \xi) - a(\varphi_k, \eta)|^2 + |b(\varphi_k, \xi) - b(\varphi_k, \eta)|_H^2\} &\leqslant \\ \leqslant \varlimsup_{n\to\infty} \{l \sup_{j\leqslant n} M|\varphi_j(\xi) - \varphi_j(\eta)|^2 + \varepsilon_n\} &= \\ = l \sup_n M|\varphi_n(\xi) - \varphi_n(\eta)|^2. \end{aligned}$$

Now, let M be a set of probability measures on X^2. This is a convex set and $\int |\varphi_j(x) - \varphi_j(y)|^2 \mu(dx, dy)$ is a nonnegative linear functional, continuous in the weak convergence topology. In this topology, M is a compactum. Inasmuch as

$$\sup_{j\leqslant n} \int (\varphi_j(x) - \varphi_j(y))^2 \mu(dx, dy)$$

increases with n and is continuous with respect to μ in the weak convergence topology, then, as $n \to \infty$, the limit

$$\sup_j \int |\varphi_j(x) - \varphi_j(y)|^2 \mu(dx, dy)$$

is a lower semicontinuous (with respect to μ) function. If it turns out to be continuous, then

$$\sup_{j\leq n} \int [\varphi_j(x) - \varphi_j(y)]^2 \mu(dx, dy) \to \sup \int [\varphi_j(x) - \varphi_j(y)]^2 \mu(dx, dy)$$

converges uniformly with respect to μ by virtue of Dini's theorem, and therefore

$$\varepsilon_n = \sup_{\mu\in M}\left(\sup_j \int [\varphi_j(x) - \varphi_j(y)]^2 \mu(dx, dy) - \sup_{j\leq n} \int [\varphi_j(x) - \varphi_j(y)]^2 \mu(dx, dy)\right) \downarrow 0.$$

Thus, a necessary and sufficient condition in order that the conditions of the existence theorem and of the uniqueness theorem coincide in the case of countable $D_0 = \{\varphi_n, n \geq 1\}$ is that the function

$$\Phi(\mu) = \sup_n \int [\varphi_n(x) - \varphi_n(y)]^2 \mu(dx, dy)$$

is continuous with respect to μ. To this end, it suffices, for example, that the family of functions $\{\varphi_n\}$ be equicontinuous.

The existence and uniqueness theorems carry over in an obvious way to equations with coefficients depending on time. We shall assume that D is some subset of C_X, separating points of X, and that there are given functions $a_s(\varphi, x)$, $b_s(\varphi, x)$ for all $s \geq 0$, satisfying the conditions of the existence theorem (uniqueness theorem). If these functions are continuous with respect to s, x for fixed $\varphi \in D$, then the equation

$$d\varphi(x_t) = a_t(\varphi, x_t)\,dt + (b_t(\varphi, x_t), dw_t)_H, \quad \varphi(x_0) = \varphi(x), \quad \varphi \in D$$

has a (unique) solution. This solution will be an X-valued Markov process.

4. Limit Theorems for Solutions of Stochastic Equations

The existence of a solution of a stochastic equation was proved in the preceding section with the aid of passage to the limit from equations for which the solution is known to exist. The passage to the limit allows one to construct solutions under much more general assumptions on the coefficients of the equation. Moreover, passage to the limit in a stochastic differential equation - i.e. the investigation of the nature of the continuous dependence of the solution of the equation on the coefficients - is of interest in itself.

We shall first establish the conditions for the compactness for a family of measures corresponding to solutions of stochastic differential equations. Let X be a compactum, let $C_{[0,T]}(X)$ be the space of X-valued continuous functions defined on [0,T]. We metrize this space, setting

$$\rho(x(\cdot),\ y(\cdot))=\sup_{t\leqslant T} r(x(t),\ y(t)),$$

where r is the metric in X. Furthermore, let Θ be an arbitrary parameter set and for $\theta \in \Theta$ let there be defined measures μ^θ on $C_{[0,T]}(X)$. Inasmuch as $C_{[0,T]}(X)$ is a complete metric space, then there is defined the concept of weak convergence for the measures μ^θ and consequently one can speak of the weak compactness of the family of measures $\{\mu^\theta,\ \theta \in \Theta\}$. General conditions for the weak compactness of a family of measures on a compact metric separable space are well known. In conformity with the space $C_{[0,T]}(X)$, they have the following form:

<u>Condition for weak compactness in</u> $C_{[0,T]}(X)$. For each $\delta > 0$,

$$\lim_{h\to 0}\sup_{\theta}\mu^\theta\{x(\cdot):\ \sup_{|t_1-t_2|\leqslant h} r(x(t_1),\ x(t_2))>\delta\}=0. \tag{1}$$

Suppose the measures μ^θ correspond to solutions of the stochastic differential equation

$$d\varphi(x_t^\theta)=a^\theta(\varphi,\ x_t^\theta)\,dt+(b^\theta(\varphi,\ x_t^\theta),\ dw_t)_H,\ \varphi\in D^\theta,\ x_0^\theta=x^\theta, \tag{2}$$

where the functions $a^\theta(\varphi, x)$ and $b^\theta(\varphi, x)$ are defined on $D^\theta \times X$, take on values in R and H respectively, linear with respect to φ and such that equation (2) has a solution, $D^\theta \subset C_X$ and the functions from D^θ separate points. For each Borel set F from $C_{[0,T]}(X)$, we set

$$\mu^\theta(F)=\mathsf{P}\{x^\theta\in F\}.$$

We shall find conditions for weak compactness for measures of the indicated form.

THEOREM 1. Suppose that, for each $\theta \in \Theta$, one can find a sequence of functions $\{\varphi_k^\theta \in D^\theta\}$ such that

(1) $\sup_\theta \sup_x \{|a^\theta(\varphi_k^\theta, x)| + |b^\theta(\varphi_k^\theta, x)|_H\} < \infty$

for all k;

(2) if $\delta_n(\theta, x, \varepsilon)$ is the diameter of the set

$$\{y: |\varphi_k^\theta(y) - \varphi_k^\theta(x)| \leqslant \varepsilon, \quad k=1,\ldots,n\},$$

then, for all $\varepsilon > 0$,

$$\lim_{\varepsilon\to 0} \overline{\lim_{n\to\infty}} \sup_{\theta\in\Theta} \sup_{x\in X} \delta_n(\theta, x, \varepsilon) = 0.$$

Then the family of measures $\{\mu^\theta, \theta \in \Theta\}$, corresponding to solutions of equation (2), is compact.

Proof. We set

$$\eta_k^\theta(t) = \int_0^t a^\theta(\varphi_k^\theta, x_s^\theta)\, ds + \int_0^t (b_\theta(\varphi_k^\theta, x_s^\theta), dw(s))_H.$$

Let

$$\gamma_k = \sup_\theta \sup_x \{|a^\theta(\varphi_k^\theta, x)| + |b^\theta(\varphi_k, x)|_H\}.$$

For all $s_0 \in [0,T]$ and $h < \rho/\gamma_k$,

$$\mathsf{P}\{\sup_{s_0\leqslant t\leqslant s_0+h} |\eta_k^\theta(t) - \eta_k^\theta(s_0)| > \rho\} \leqslant$$

$$\leqslant \mathsf{P}\left\{\sup_{s_0\leqslant t\leqslant s_0+h} \left|\int_{s_0}^t (b^\theta(\varphi_k^\theta, x_s^\theta), dw(s))_H\right| > \rho - h\gamma_k\right\} \leqslant$$

$$\leqslant \mathsf{M}\left(\int_{s_0}^{s_0+h} (b^\theta(\varphi_k^\theta, x_s^\theta), dw(s)\right)^4 \cdot \left(\frac{4}{3}\right)^4 (\rho - h\gamma_k)^{-4}$$

(we have made use of well-known representations for martingales; see [15, Vol. 1, p.78, Cor.]).

Using the fact that

$$\int_{s_0}^t (b^\theta(\varphi_k^\theta, x_s^\theta), dw(s))_H = \int_{s_0}^t |b^\theta(\varphi_k^\theta, x_s^\theta)|_H\, dw_1(s),$$

where

$$w_1(t)-w_1(s_0)=\int_{s_0}^{t}\frac{1}{|b^\theta(\varphi_k^\theta,\ x_s^\theta)|_H}(b^\theta(\varphi_k^\theta,\ x_s^\theta),\ dw(s))_H$$

and therefore $w_1(t) - w_1(s_0)$ is a Wiener process in R, we convince ourselves on the basis of a well-known inequality for one-dimensional stochastic integrals (see [31, Ch. 2, §2, Th. 4]) that

$$\mathsf{M}\left[\int_{s_0}^{s_0+h}(b^\theta(\varphi_k^\theta,\ x_s^\theta),\ dw(s))_H\right]^4\leqslant 36\gamma_k^4h^2.$$

Consequently,

$$\mathsf{P}\{\sup_{s_0\leqslant t\leqslant s_0+h}|\eta_k^\theta(t)-\eta_k^\theta(s_0)|>\rho\}\leqslant\frac{4^4\gamma_k^4h^2}{9(\rho-h\gamma_k)^4}.$$

If there can be found t_1 and t_2 such that $|t_2 - t_1| \leq h$ and $|\eta_k^\theta(t_1) - \eta_k^\theta(t_2)| > 2\rho$, then taking an m such that $mh \leq t_1 \leq mh + 2h$ and $mh \leq t_2 \leq mh + 2h$, we shall have

$$2\rho<|\eta_k^\theta(t_1)-\eta_k^\theta(mh)|+|\eta_k^\theta(t_2)-\eta_k^\theta(mh)|,$$

so that

$$\sup_{mh\leqslant t\leqslant mh+2h}|\eta_k^\theta(t)-\eta_k^\theta(mh)|>\rho.$$

Therefore,

$$\mathsf{P}\{\sup_{|t_1-t_2|\leqslant h}|\eta_k^\theta(t_1)-\eta_k^\theta(t_2)|>2\rho\}\leqslant$$
$$\leqslant\sum_{mh<T}\mathsf{P}\{\sup_{mh\leqslant t\leqslant mh+2h}|\eta_k^\theta(t)-\eta_k^\theta(mh)|>\rho\}\leqslant$$
$$\leqslant\frac{T}{h}\cdot\frac{4^4}{9}\cdot\frac{\gamma_k^4(2h)^2}{(\rho-2h\gamma_k)^4}\leqslant\frac{4^7}{9}T\gamma_k^4\rho^{-4}h,$$

provided $h \leq \rho/4\gamma_k$. We denote the factor before h in the right member of the last inequality by $\alpha_k(2\rho)$. Suppose n and ε have been chosen so that

$$\sup_{\theta\in\Theta}\sup_{x\in X}\delta_n(\theta,\ x,\ \varepsilon)\leqslant\delta.$$

Then: if $|\varphi_k^\theta(x) - \varphi_k^\theta(y)| \leq \varepsilon$ for $k \leq n$, then $r(x, y) \leq \delta$. Consequently,

$$\mathsf{P}\{\sup_{|t_1-t_2|\leqslant h} r(x^\theta_{t_1}, x^\theta_{t_2}) > \delta\} \leqslant$$
$$\leqslant \mathsf{P}\Big\{\bigcup_{k=1}^{n}\{\sup_{|t_1-t_2|\leqslant h} |\varphi^\theta_k(x^\theta_{t_1}) - \varphi^\theta_k(x^\theta_{t_2})| > \varepsilon\}\Big\} \leqslant$$
$$\leqslant \sum_{k=1}^{n} \mathsf{P}\{\sup_{|t_1-t_2|\leqslant h} |\varphi^\theta_k(x^\theta_{t_1}) - \varphi^\theta_k(x^\theta_{t_2})| > \varepsilon\} \leqslant \sum_{k=1}^{n} \alpha_k\left(\frac{\varepsilon}{4}\right)h,$$

provided

$$h \leqslant \frac{\varepsilon}{8\max\limits_{k\leqslant n} \gamma_k}.$$

Hence, for all $\delta > 0$,

$$\lim_{h\to 0}\sup_{\theta\in\Theta} \mathsf{P}\{\sup_{|t_1-t_2|\leqslant h} r(x^\theta_{t_1}, x^\theta_{t_2}) > \delta\} = 0,$$

i.e. condition (1) is satisfied. This completes the proof of the theorem.

REMARK. We obtain an important special case of a family of stochastic differential equations when $D^\theta = D$ does not depend on θ. As noted in the preceding section, one can always find a sequence $\{\varphi_k, k = 1, 2, \ldots\}$, separating points of X. If wet set $\varphi^\theta_k = \varphi_k$, then condition (2) is automatically satisfied. Consequently, compactness of the family $\{\mu_\theta\}$ is sufficient in order that, for all k,

$$\sup_\theta \sup_x \{|a^\theta(\varphi_k, x)| + |b^\theta(\varphi_k, x)|_H\} < \infty.$$

We now consider the form of limit points for a weakly compact family of measures $\{\mu^\theta, \theta \in \Theta\}$ under certain conditions on the coefficients of the stochastic equation. Obviously, it suffices to consider only the case of countable Θ.

THEOREM 2. Let $\theta = 1, 2, \ldots$ and suppose the sequence of measures μ^θ converge weakly as $\theta \to \infty$ to some measure μ on $C_{[0,T]}(X)$, where μ corresponds to the solution of equation (2), whereby $D^\theta = D$ does not depend on θ, and, for all $\varphi \in D$, there exist functions $a(\varphi, x)$ and $b(\varphi, x)$, continuous with respect to x, such that

$$\lim_{\theta\to\infty}\sup_x \{|a(\varphi, x) - a^\theta(\varphi, x)| + |b(\varphi, x) - b^\theta(\varphi, x)|_H\} = 0.$$

Then the measure μ corresponds to the process x_t for which, for all $\varphi \in D$, the relation

$$d\varphi(x_t) = a(\varphi, x_t)\,dt + (b(\varphi, x_t), dw_t)_R.$$

is satisfied.

Proof. We shall show that for all $\varphi \in D$ the process

$$\eta_\varphi(t) = \varphi(x_t) - \int_0^t a(\varphi, x_s)\,ds$$

is a martingale relative to the filtration F_t generated by the process x_t. Let $s_1 < s_2 < \ldots < s_k \leq t_1 < t_2$, $g(x_1, \ldots, x_k)$ a numerical function continuous on X^k. It follows from the weak convergence of μ^θ to μ that

$$\lim_{\theta\to\infty} \int f(x(\cdot))\,\mu^\theta(dx) = \int f(x(\cdot))\,\mu(dx)$$

for every continuous bounded function $f(x(\cdot))$, defined on $C_{[0,T]}(X)$. Inasmuch as the function

$$\left[\varphi(x(t_2)) - \varphi(x(t_1)) - \int_{t_1}^{t_2} a(\varphi, x(s))\,ds\right] g(x(s_1), \ldots, x(s_k))$$

is continuous on $C_{[0,T]}(X)$, then

$$\begin{aligned} \mathsf{M}\,[\eta_\varphi(t_2) - \eta_\varphi(t_1)]\,g(x_{s_1}, \ldots, x_{s_k}) &= \\ = \mathsf{M}\left[\varphi(x_{t_2}) - \varphi(x_{t_1}) - \int_{t_1}^{t_2} a(\varphi, x_s)\,ds\right] g(x_{s_1}, \ldots, x_{s_k}) &= \\ = \lim_{\theta\to\infty} \mathsf{M}\left[\varphi(x^\theta_{t_2}) - \varphi(x^\theta_{t_1}) - \int_{t_1}^{t_2} a(\varphi, x^\theta_s)\,ds\right] g(x^\theta_{s_1}, \ldots, x^\theta_{s_k}). \end{aligned}$$

It follows from equation (2) that

$$\varphi(x^\theta_t) - \int_0^t a^\theta(\varphi, x_s)\,ds$$

is a martingale. Consequently,

$$\mathsf{M}\left[\varphi(x^\theta_{t_2}) - \varphi(x^\theta_{t_1}) - \int_{t_1}^{t_2} a^\theta(\varphi, x^\theta_s)\,ds\right] g(x^\theta_{s_1}, \ldots, x^\theta_{s_k}) = 0.$$

Inasmuch as

$$\lim_{\theta\to\infty} \mathsf{M}\int_{t_1}^{t_2} |a(\varphi, x_s^\theta) - a^\theta(\varphi, x_s^\theta)|\, ds \leqslant$$
$$\leqslant (t_2 - t_1) \lim_{\theta\to\infty} \sup_x |a(\varphi, x) - a^\theta(\varphi, x)| = 0,$$

then

$$\mathsf{M}[\eta_\varphi(t_2) - \eta_\varphi(t_1)]\, g(x_{s_1}, \ldots, x_{s_k}) = 0$$

for arbitrary k, $s_1 < \ldots < s_k \leqq t_1 < t_2$ and continuous function on X^k. Hence,

$$\mathsf{M}[\eta_\varphi(t_2) - \eta_\varphi(t_1)/\mathscr{F}_{t_1}] = 0.$$

We shall show that the joint characteristic of the martingales $\eta_{\varphi_1}(t)$ and $\eta_{\varphi_2}(t)$ equals

$$\int_0^t (b(\varphi_1, x_s), b(\varphi_2, x_s))_H\, ds. \tag{5}$$

For this, it suffices to show that the process

$$\eta_{\varphi_1}(t)\, \eta_{\varphi_2}(t) - \int_0^t (b(\varphi_1, x_s), b(\varphi_2, x_s))_H\, ds,$$

is a martingale, i.e. that, for $s_1 < s_2 < \ldots < s_k \leqq t_1 < t_2$

$$\mathsf{M}\Big[\eta_{\varphi_1}(t_2)\, \eta_{\varphi_2}(t_2) - \eta_{\varphi_1}(t_1)\, \eta_{\varphi_1}(t_1) -$$
$$- \int_{t_1}^{t_2} (b(\varphi_1, x_s), b(\varphi_2, x_s))_H\, ds\Big]\, g(x_{s_1}, \ldots, x_{s_k}) = 0.$$

For this, we make use of the continuity of the functional

$$\Big[(\varphi_1(x(t_2)) - \int_0^{t_2} a(\varphi_1, x(s))\, ds)(\varphi_2(x(t_2)) -$$
$$- \int_0^{t_2} a(\varphi_2, x(s))\, ds) - (\varphi_1(x(t_1)) -$$
$$- \int_0^{t_1} a(\varphi_1, x(s))\, ds)(\varphi_2(x(t_1)) - \int_0^{t_1} a(\varphi_2, x(s))\, ds) -$$
$$- \int_{t_1}^{t_2} (b(\varphi_1, x(s)), b(\varphi_2, x(s)))_H\, ds\Big]\, g(x(s_1), \ldots, x(s_k))$$

on $C_{[0,T]}(X)$ and the fact that, by virtue of equation (2),

$$(\varphi_1(x_t^\theta) - \int_0^t a^\theta(\varphi_1, x_s^\theta)\,ds)(\varphi_2(x_t^\theta) - \int_0^t a^\theta(\varphi_2, x_s^\theta)\,ds) - \\ - \int_0^t (b^\theta(\varphi_1, x_s^\theta), b(\varphi_2, x_s^\theta))_H\,ds$$

is a martingale. Therefore,

$$\mathsf{M}\Big[\eta_{\varphi_1}(t_2)\eta_{\varphi_2}(t_2) - \eta_{\varphi_1}(t_1)\eta_{\varphi_2}(t_1) - \\ - \int_{t_1}^{t_2} (b(\varphi_1, x_s), b(\varphi_2, x_s))_H\,ds\Big] g(x_{s_1}, \ldots, x_{s_k}) = \\ = \lim_{\theta\to\infty} \mathsf{M}\Big[\varphi_1(x_{t_2}^\theta)\int_0^{t_2}(a^\theta(\varphi_2, x_s^\theta) - a(\varphi_2, x_s^\theta))\,ds - \\ - \varphi_1(x_{t_1}^\theta)\int_0^{t_1}(a^\theta(\varphi_2, x_s^\theta) - a(\varphi_2, x_s^\theta))\,ds + \\ + \varphi_2(x_{t_2}^\theta)\int_0^{t_2}(a^\theta(\varphi_1, x_s^\theta) - a(\varphi_1, x_s^\theta))\,ds - \\ - \varphi_2(x_{t_1}^\theta)\int_0^{t_1}(a^\theta(\varphi_1, x_s^\theta) - a(\varphi_1, x_s^\theta))\,ds + \\ + \int_0^{t_2} a(\varphi_1, x_s^\theta)\,ds \int_0^{t_2} a(\varphi_2, x_s^\theta)\,ds - \int_0^{t_2} a^\theta(\varphi_1, x_s^\theta)\,ds \int_0^{t_2} a^\theta(\varphi_2, x_s^\theta)\,ds - \\ - \int_0^{t_1} a(\varphi_1, x_s^\theta)\,ds \int_0^{t_1} a(\varphi_2, x_s^\theta)\,ds + \int_0^{t_1} a^\theta(\varphi_1, x_s^\theta)\,ds \int_0^{t_1} a^\theta(\varphi_2, x_s^\theta)\,ds + \\ + \int_{t_1}^{t_2} [(b^\theta(\varphi_1, x_s^\theta), b^\theta(\varphi_2, x_s^\theta))_H - (b(\varphi_1, x_s^\theta), b(\varphi_2, x_s^\theta))_H]\,ds\Big] \times \\ \times g(x_{s_1}, \ldots, x_{s_k}) = O(\overline{\lim_{\theta\to\infty}} \sup_x \{|a(\varphi_1, x) - a^\theta(\varphi_1, x)| + \\ + |a(\varphi_2, x) - a^\theta(\varphi_2, x)| + |b(\varphi_1, x) - b^\theta(\varphi_1, x)|_H + \\ + |b(\varphi_2, x) - b^\theta(\varphi_2, x)|_H\}) \to 0.$$

Thus, we established that the $\eta_\varphi(t)$ are martingales relative to the filtration F_t and that the joint characteristic of their characteristics $\eta_{\varphi_1}(t)$ and $\eta_{\varphi_2}(t)$ is given by expression (5). Using the proof of Theorem 1, §2 (the topology in $\Phi = D$ was needed only for the proof of the factored representation (4), §2, which in our case exists by virtue of formula (5)), we convince ourselves of the existence of a Wiener process $w(t)$ in H such that formula (4) is valid. (We note that inasmuch as we are dealing with measures, the question of how to denote the indicated process has no significance, therefore it can simply be denoted as the process that figures in equation (2). To be precise, it would be necessary, denoting it by $\tilde{w}$, to construct a new process x such that the joint distribution of $\tilde{x}$ and w coincide with the joint distribution of x and $\tilde{w}$. This process $\tilde{x}$ will also satisfy relation (2). Obviously, to the processes x and $\tilde{x}$ there corresponds one and the same measure μ.) This completes the proof of the theorem.

REMARK. One can make use of Theorem 2 for the proof of the existence of solutions of a stochastic differential equation. We shall assume that for the coefficients $a(\varphi, x)$ and $b(\varphi, x)$ one can find sequences $a^\theta(\varphi, x)$ and $b^\theta(\varphi, x), \theta = 1, 2, \ldots$, such that equation (2) has a solution, whereby $D^\theta = D$ does not depend on θ. If for some countable set $D_0 \subset D$ condition (4) is satisfied, then the conditions of Theorem 1 are satisfied and hence the family of measures $\{\mu^\theta, \theta = 1, 2, \ldots\}$ is compact. Therefore, the sequence μ^θ has a convergent subsequent μ^{θ_k}. For this sequence the conditions of Theorem 2 are satisfied. Hence, the limit measure μ will correspond to the solution of equation (4) for $\varphi \in D_0$. If D_0 is such that for all $\varphi \in D$ one can find a sequence $\varphi_n \in D_0$ such that $\varphi_n(x)$, $a(\varphi_n, x)$, $b(\varphi_n, x)$ are uniformly bounded and converge respectively to $\varphi(x)$, $a(\varphi, x)$, $b(\varphi, x)$, then passing to the limit in the equation

$$\varphi_n(x_t) - \varphi_n(x_0) = \int_0^t a(\varphi_n, x_s)\,ds + \int_0^t (b(\varphi_n, x_s), dw_s)_H$$

we convince ourselves that equation (4) is valid for all $\varphi \in D$.

More interesting results are obtained if the processes x_t^θ are given, generally speaking, on different sets. In order that it be possible to speak of the convergence of processes, we shall assume that each process is given on some closed subset X^θ of a compact space X. Functions from D^θ are given on X^θ and equation (2) is satisfied for the processes x_t^θ. Inasmuch as all processes ahve phase space X, then Theorem 1 is applicable to them with one, natural modification: the sequence $\{\varphi_k^\theta\}$ separate points of X^θ.

THEOREM 3. Let $\Theta = \{1, 2, \ldots\}$ and suppose x_t^θ is a solution of equation (2), where D^θ consists of functions defined on X^θ. We shall assume that there exists a closed subset $X^0 \subset X$, a sequence of continuous functions $S^\theta(x)$, defined on X^θ and taking on values in X^0, for which $x \in X^0$,

$r(S^\theta(x), x) \to 0$ as $\theta \to \infty$, and also, for all θ, the functions $\{\varphi_k^\theta, \theta = 1, 2, \ldots\}$ from D^θ, for which the following conditions are satisfied:

(a) $$\lim_{\varepsilon \to 0} \limsup_{n \to \infty} \sup_{\theta \in \Theta} \sup_{x \in X^\theta} \delta_n(\theta, x, \varepsilon) = 0,$$

(b) $$\sup_{\theta \in \Theta} \sup_{x \in X^\theta} \{|\varphi_k^\theta(x)| + |a^\theta(\varphi_k^\theta(x)| + |b^\theta(\varphi_k^\theta, x)|_H\} < \infty.$$

Then the family of measures $\{\mu^\theta, \theta \in \Theta\}$, where μ^θ corresponds to x^θ on $C_{[0,T]}(X)$, is compact.

If there exist sequences of continuous functions $\{\varphi_k(x)\}$, $\{a_k(x)\}$, $\{b_k(x)\}$, defined on X^0, taking on values in R, R and H respectively such that

$$\lim_{\theta \to \infty} \sup_{x \in X^\theta} \{|\varphi_k^\theta(x) - \varphi_k(S^\theta(x))| + |a^\theta(\varphi_k^\theta, x) - a_k(S^\theta(x))| + $$
$$+ |b^\theta(\varphi_k^\theta, x) - b_k(S^\theta x)|_H\} = 0,$$

then every limit measure μ of the family $\{\mu^\theta, \theta \in \Theta\}$ corresponds to a random process x_t for which

$$d\varphi_k(x_t) = a_k(x_t)\,dt + (b_k(x_t), dw_t)_H, \quad k = 1, 2, \ldots \tag{6}$$

Proof. The compactness of the family $\{\mu^\theta, \theta \in \Theta\}$ follows from Theorem 1. Relation (6) is proved in the same manner as relation (4) in Theorem 2. We observe the difference in the proof, connected with the fact that the functions φ_k^θ now depend on θ and are defined on different sets, by proving that $\eta_k(t) = \varphi_k(x_t) - \int_0^t a_k(x_s)ds$ is a martingale.

Inasmuch as $\varphi_k^\theta(x_t^\theta) - \int_0^t a^\theta(\varphi_k^\theta, x_s)ds$ is a martingale, then

$$\mathsf{M}\left[\varphi_k^\theta(x_{t_2}^\theta) - \varphi_k^\theta(x_{t_1}^\theta) - \int_{t_1}^{t_2} a^\theta(\varphi_k^\theta, x_s^\theta)\,ds\right] g(x_{s_1}^\theta, \ldots, x_{s_m}^\theta) = 0,$$

for arbitrary m, $s_1 < \ldots < s_m \leq t_1 < t_2$ and the function $g(x_1, \ldots, x_m)$ is continuous on X^m. Therefore

$$\lim_{\theta \to \infty} \mathsf{M}\left[\varphi_k(S^\theta(x_{t_2}^\theta)) - \varphi_k(S^\theta(x_{t_1}^\theta)) - \right.$$
$$\left. - \int_{t_1}^{t_2} a_k(S^\theta(x_s^\theta))\,ds\right] g(S^\theta(x_{s_1}^\theta), \ldots, S^\theta(x_{s_m}^\theta)) = 0. \tag{7}$$

It remains to note that the sequence of processes $S^\theta(x_t^\theta)$ also weakly converges to the process x_t since, for an arbitrary continuous bounded function $f(x(\cdot))$ on $C_{[0,T]}(X)$,

$$\lim_{\theta\to\infty} \mathsf{M}\,|\,f(x^\theta(\cdot)) - f(S^\theta(x^\theta(\cdot)))\,| = 0.$$

Therefore, the limit in the left member of (7) equals

$$\mathsf{M}\left[\varphi_k(x_{t_2}) - \varphi_k(x_{t_1}) - \int_{t_1}^{t_2} a_k(x_s)\,ds\right] g(x_{s_1}, \ldots, x_{s_m})$$

and hence $\eta_k(t)$ is a martingale. It is established analogously that the joint characteristic of the martingales $\eta_k(t)$ and $\eta_i(t)$ equals

$$\int_0^t (b_k(x_s),\ b_i(x_s))_H\,ds.$$

This completes the proof of the theorem.

REMARK. As Theorem 2, Theorem 3 can be used in the proof of the existence of a solution of a stochastic equation (cf. Remark to Theorem 2). In this connection, the fact is essential that the compactum X can always be approximated by 'finite-dimensional' compacta. We shall clarify the meaning of the preceding assertion. First of all, X can be embedded isometrically into a linear space of bounded numerical sequences $\{\alpha_1, \alpha_2, \ldots\}$ with norm

$$\|\{\alpha_1, \alpha_2, \ldots\}\| = \sup_k |\alpha_k|.$$

This embedding is realized by the function $x \to \{r(x, x_1), r(x, x_2), \ldots\}$, where $\{x_k\}$ is a sequence that is dense in X. In fact,

$$r(x, y) = \sup_k |r(x, x_k) - r(y, x_k)|.$$

Therefore, X can at once be assumed to be a compact set of some separable Banach space (the closure of the linear hull of the image of X under the indicated embedding will be such). Taking as X^θ the intersection of X with the linear hull of a finite $(1/\theta)$-net in X, we obtain a compactum lying in a finite-dimensional subspace - it is natural to assume it to be finite-dimensional.

COROLLARY. Suppose X is a compactum in some Hilbert space and that the set of functions D is such that, for $\varphi \in D$ and $\varphi(Px) \in D$, where P is the projection operator on some finite-dimensional subspace. If there exists a sequence of finite-dimensional projectors P_n such that $P_n x \to x$ and

$$\lim_{n\to\infty}\sup_{x\in X}\{|a(\varphi(P_n), P_nx)-a(\varphi, x)|+ \\ +|b(\varphi(P_n), P_nx)-b(\varphi, x)|_H\}=0$$

and the stochastic equation

$$d\varphi(P_nx_t^n)=a(\varphi(P_n), P_nx_t^n)+(b(\varphi(P_n), P_nx_t^n), dw_t)_H$$

has a solution on the set $X^n = P_nX$, then equation (4) has a solution.

5. Weak Solutions

The solutions of a stochastic equation under the conditions of the uniqueness theorem (see §3) possess the following important property: they are measurable relative to the filtration of the σ-algebras F_t, generated by the Wiener process appearing in the equation. Such solutions are called strong solutions. If a solution is not necessarily measurable relative to F_t (the initial value is assumed to be nonrandom), then it is called a weak solution; in this case, it exists in some extension of a probability space which can be obtained by adjoining to the Wiener process w_t an independent uniformly distributed quantity. The preceding assertion was in essence established in §2, where it was shown - with reference to a family of distributions - how to construct a solution of the stochastic equation and the Wiener process; in this connection it would have been impossible to assert that the solution is uniquely determined by a Wiener process, so that the solutions were considered as weak ones. In general, giving a weak solution is equivalent to giving the joint distribution of the processes x_t and w_t. If the weak solution x_t is constructed on a probability space generated by the process w_t and is uniformly distributed by the variable ξ, then there exists a function $g_t(s, z(\cdot))$, $s \in [0, 1]$, such that

$$x_t=g_t(\xi, w(\cdot)).$$

For every other uniformly distributed variable $\hat{\xi}$ the process $\hat{x}_t = g_t(\hat{\xi}, w(\circ))$ will also be a solution. Therefore, on the probability space generated by $w(\cdot)$, ξ and $\hat{\xi}$, there will be two solutions provided $g_t(s, z(\cdot))$ really depends on s. If the solution is unique, then $g_t(s, w(\circ))$ does not depend on s, and hence x_t is a function of $w(\circ)$ only, i.e. it is a strong solution (see [11, Vol. 3, p.340]).

For weak solutions, one introduces the concept of weak uniqueness. A solution of the equation

$$d\varphi(x_t)=a(\varphi, x_t)\,dt+(b(\varphi, x_t), dw(t))_H,\quad \varphi\in D,\ x_0=x \tag{1}$$

is called a weakly unique solution if for any solutions x_t and $\bar{x}_t$ of this equation the measures corresponding to them on $C_{[0,\infty)}(X)$ coincide.

THEOREM 1. Suppose X is a locally compact space, D is everywhere dense in the space C_0 of functions $\varphi(x)$ for which $\varphi(x) \to 0$ as $x \to \infty$, $a(\varphi, x) \in C_0$, $(b(\varphi, x), h)_H \in C_0$ for $\varphi \in C_0$, $h \in H$. A necessary and sufficient condition for a solution of equation (1) to be weakly unique for an arbitrary initial value $x \in X$ is that there exists a semigroup of operators T_t on C_0 such that $||T_t||_H$ is locally bounded and, for all $\varphi \in D$ $T_t\varphi \in D$ and

$$\frac{d}{dt} T_t\varphi(x) = T_t a(\varphi, x) \tag{2}$$

(i.e. D lies in the domain D_A of the infinitesimal generator A of the semigroup T_t and $A\varphi(x) = a(\varphi, x)$). If X is a compactum and the solution of (1) exists and is weakly unique, the x_t is a homogeneous Markov process, and the semigroup of operators corresponding to it satisfies relation (2).

Proof. Suppose x_t is a solution of (1). Then

$$d_s T_{t-s}\varphi(x_s) = \left[\frac{d}{ds} T_{t-s}\varphi(y)\right]_{y=x(s)} ds + T_{t-s}[a(\varphi, x_s)]\,ds + \\ + T_{t-s}[(b(\varphi, y), dw_s)_H]_{y=x_s} \tag{3}$$

and hence $\mathsf{M} d_s T_{t-s}\varphi(x_s) = 0$. Therefore

$$\mathsf{M}[T_t\varphi(x) - \varphi(x_t)] = 0, \quad \mathsf{M}\varphi(x_t) = T_t\varphi(x).$$

Inasmuch as the preceding relation is valid on D and D is dense in C_0, then it is also valid for all $\varphi \in C_0$, in particular for $f \in C_0$ $\mathsf{M}f(x_t)$ as a function of the initial value belongs to C_0. Let $u > 0$. Using the equation

$$T_t\varphi(x_u) - \varphi(x_{t+u}) = \int_0^t T_{t-s}[(b(\varphi, y), dw_{s+u})]_{y=x_{s+u}},$$

which is obtained from (3) if one replaces there x_s by x_{s+u}, we obtain

$$\mathsf{M}[\varphi(x_{t+u})/x_v, \ v \leqslant u] = T_t\varphi(x_u), \qquad \varphi \in D,$$

hence, for all $f \in C_0$

$$\mathsf{M}[f(x_{t+u})/x_v, \ v \leqslant u] = T_t f(x_u). \tag{4}$$

Multiplying this relation by $f_1(x_u)$, where $f_1 \in C_o$, and taking the mathematical expectation and taking into consideration that $f_1(x)T_tf(x) \in C_o$, we find that

$$\mathsf{M}f_1(x_u)f(x_{t+u}) = T_u[f_1(x)T_tf(x)].$$

We find analogously that for arbitrary $t_1 < t_2 < \ldots < t_n$, $f_1, f_2, \ldots, f_n \in C_o$,

$$\mathsf{M}f_1(x_{t_1})f_2(x_{t_2})\ldots f_n(x_{t_n}) = \\ = T_{t_1}[f_1(x)T_{t_2-t_1}[f_2(x)\ldots T_{t_n-t_{n-1}}f_n(x)]]. \tag{5}$$

Obviously, the mathematical expectations in the left member of (5) define uniquely the measure corresponding to the process x_t. This proves the first assertion of the theorem.

To prove the second assertion, we consider the function $S_t(\varphi, x)$, defined for all $\varphi \in C_o$, $t > 0$, $x \in X$, by the equation

$$S_t(\varphi, x) = \mathsf{M}\varphi(x_t), \tag{6}$$

where x_t is the solution of (1) with initial condition $x_o = x$. We assume that $x_n = x$. Using Theorem 1, §4, it is easy to verify that the sequence of measures μ_{x_n}, corresponding to equation (1) with initial condition $x(0) = x_n$, is compact. On the basis of Theorem 2, §4, every limit point of this sequence is a measure corresponding to some solution of equation (1) with initial condition $x(0) = x$. By virtue of the weak uniqueness, this measure is unique. Thus, the sequence of measures $\{\mu_{x_n}\}$ is weakly compact and has a unique limit point μ_x, corresponding to the solution of (1) with initial condition $x_o = x$. Therefore, μ_{x_n} converges weakly to μ_x. In particular, the function $S_t(\varphi, x)$ is continuous with respect to x. We consider the solution $\hat{x}_t$ of equation (1) with random initial condition $x(0) = \eta$, where η does not depend on the Wiener process w_t. Let $\hat{\mu}$ be the measure corresponding to the solution and let $\hat{\mu}(A/\eta)$ be the conditional measure (the probability given the initial value). It is easy to verify that for each countable set $D_o \subset D$

$$\mathsf{P}\left\{\varphi(\hat{x}_t) - \varphi(\eta) = \int_0^t a(\varphi, \hat{x}_s)\,ds + \right. \\ \left. + \int_0^t (b(\varphi, \hat{x}_s), dw_s)_H, \quad \varphi \in D_0,\ t \geqslant 0/\eta\right\} = 1$$

with probability 1. Therefore, $\hat{\mu}(A/\eta)$ corresponds to the solution of (1) for almost all η if D is replaced by D_0. Choosing D_0 so that it is dense in C_0 and dense in D relative to the topology in which $a(\varphi, \circ)$ and $b(\varphi, \cdot)$ are continuous, we convince ourselves that $\hat{\mu}(A/\eta)$ in fact corresponds to the solution of (1) for almost all η if $x = \eta$, i.e. $\hat{\mu}(A/\eta) = \mu_\eta$ with probability 1.

We consider the solution of (1) at the instant $t + s$, $t > 0$, $s > 0$, representing it in the following way:

$$\varphi(x_{t+s}) - \varphi(x_s) = \\ = \int_0^t a(\varphi, x_{s+u})\,du + \int_0^t (b(\varphi, x_{s+u}), d(w_{s+u} - w_s))_H.$$

Since $w_{s+u} - w_s$ is a Wiener process with respect to u which does not depend on x_s, then, by our statement above, for each collection $t_1 < t_2 < \ldots < t_n$ and functions $f_1(x), \ldots, f_n(x) \in C_0$,

$$\mathsf{M}\left[\prod_{k=1}^{n} f_k(x_{s+t_k})/x_s\right] = \int \left(\prod_{k=1}^{n} f_k(x_{t_k})\right) \mu_{x_s}(dx(\cdot)).$$

Hence,

$$\mathsf{M} f(x_s) \prod_{k=1}^{n} f_k(x_{s+t_k}) = S_s\left(f(\cdot) \int \prod_{k=1}^{n} f_k(x_{t_k})\,\mu_\cdot(dx(\cdot)),\ x\right) = \\ = \int f(x_s) \prod_{k=1}^{n} f_k(x_{s+t_k})\,\mu_x(dx(\cdot)).$$

It follows from formula (7) that for $t_1 < \ldots < t_{n+1}$, $f_1, \ldots, f_{n+1} \in C_0$,

$$\int \prod_{k=1}^{n+1} f_k(x_{t_k})\,\mu_x(dx(\cdot)) = \\ = S_{t_1}(f_1(\cdot)\,S_{t_2-t_1}(f_2(\cdot) \ldots S_{t_{n+1}-t_n}(f_{n+1}, \cdot) \ldots), x),$$

whence

$$\mathsf{M} f_1(x_{t_1}) \ldots f_{n+1}(x_{t_{n+1}}) = \mathsf{M} f_1(x_{t_1}) \ldots f_n(x_{t_n})\,S_{t_{n+1}-t_n}(f_{n+1}, x_{t_n}).$$

Hence

$$\mathsf{M}(f(x_{t+u})/x_v,\ v \leqslant u) = S_t(f, x_u)$$

with probability 1. Therefore, x_t is a homogeneous Markov process and its transition probability $P(t, x, dy)$ satisfies the relation

$$\int f(y)\, P(t, x, dy) = S_t(f, x),$$

i.e. $S_t(f, x) = T_t f(x)$, where T_t is the semigroup corresponding to this Markov process. For

$$\varphi \in D \quad T_t\varphi(x) - \varphi(x) = \int_0^t \mathsf{M}a(\varphi, x_s)\, ds,$$

where x_s is the solution of (1) with initial condition $x(0) = x$.

It follows from the continuity of $a(\varphi, x)$ and the process x_s that

$$\frac{d}{dt} T_t\varphi(x) = \mathsf{M}a(\varphi, x_t) = T_t a(\varphi, x),$$

i.e. (2) is satisfied. This completes the proof of the theorem.

We shall now study the question of the absolute continuity of measures corresponding to solutions of stochastic differential equations with different coefficients. In this connection, we construct some class of transformations of random processes that carry weak solutions of one stochastic differential equation into the solution of another. By the same token we shall obtain a new method for constructing a weak solution - the method of the absolutely continuous change of measure.

Let x_t be a solution of equation (1). We shall assume that the process x_t is defined on some probability space $\{\Omega, \mathfrak{S}, \mathsf{P}\}$ on which is given the Wiener process w_t occurring in the equation. We consider another probability space, differing from the initial one only by the measure, $\{\Omega, \mathfrak{S}, \tilde{\mathsf{P}}\}$, where

$$\tilde{\mathsf{P}}(A) = \int_A \rho(\omega)\, \mathsf{P}(d\omega),$$

where $\rho(\omega) \geq 0$, $\int \rho(\omega)\mathsf{P}(d\omega) = 1$. On the probability space $\{\Omega, \mathfrak{S}, \tilde{\mathsf{P}}\}$, the process x_t will have another distribution. We shall be interested in the case when x_t will also satisfy an equation of the form (1) with some other coefficients and another Wiener process on the new probability space. We shall denote by $\mathcal{F}_t$ the flow of σ-algebras generated by the process x_t on the initial probability space,

$$\rho_t(\omega) = \mathsf{M}(\rho(\omega)/\mathcal{F}_t),$$

where $\rho_t(\omega)$ is a nonnegative martingale, whereby, for every $\mathcal{F}_t$-measurable variable ξ,

$$\tilde{M}\xi \doteq \int \xi(\omega)\tilde{P}(d\omega) = M\rho_t(\omega)\xi.$$

We denote by $J(P)$ the collection of processes $\xi(t)$ adapted to F_t and having an Itô differential:

$$d\xi(t) = \alpha(t)\,dt + d\beta_t,$$

where $\alpha(t)$ is an integrable (with respect to t) F_t-measurable process, and β_t is an F_t-martingale with absolutely continuous characteristic. Obviously, in our case $\varphi(x_t) \in J(P)$ for all $\varphi \in D$. If x_t also satisfies an equation of type (1) on the probability space $\{\Omega, \mathfrak{S}, \tilde{P}\}$, $\varphi(x_t) \in J(\tilde{P})$ for $\varphi \in D$. This will be fulfilled, for example, if $J(P) \subset J(\tilde{P})$.

We shall assume that the martingale $\rho_t(\omega)$ is representable in the form

$$\rho_t(\omega) = 1 + \int_0^t (c_s(\omega),\ dw_s)_H, \tag{8}$$

where $c_s(\omega)$ is an F_s-measurable H-valued function. (In the case when the σ-algebra F_t is generated by a Wiener process - i.e. for a strong solution x_t formula (8) always holds - that follows from the form of F_t-measurable quantities (see [11, Vol. 3, p.320]).) We shall need the following result of N.V. Girsanov (see [11, Vol. 3, p.330]).

LEMMA. If $\rho_t(\omega)$ and formula (8) is valid where $|c_s(\omega)|_H^2\rho_s^{-2}(\omega)$ is a function that is locally integrable with respect to s, then $J(P) \subset J(\tilde{P})$ and for $\eta(t) \in J(P)$, for which

$$d\eta(t) = \alpha_t(\omega)\,dt + (b_t(\omega),\ dw_t)_H,$$

the relation

$$d\eta(t) = [\alpha_t(\omega) + (b_t(\omega),\ c_t(\omega))_H\rho_t^{-1}(\omega)]\,dt + (b_t(\omega),\ d\tilde{w}_t)_H, \tag{9}$$

holds, where

$$\tilde{w}(t) = w(t) - \int_0^t \rho_s^{-1}(\omega)c_s(\omega)\,ds$$

is a Wiener process on $\{\Omega, \mathfrak{S}, \tilde{P}\}$.

In the proof use is made only of the fact that $\tilde{w}_t$ is a Wiener process on $\{\Omega, \mathfrak{S}, \tilde{P}\}$. We consider $(\tilde{w}_t, z)_H$. We shall show that this is a local martingale. To this end, it suffices to show that $\rho_t(\omega)(\tilde{w}_t, z)_H$ will be a local martingale on $\{\Omega, \mathfrak{F}, P\}$. But

$$d[\rho_t(\omega)(\tilde{w}(t), z)_H] = (\tilde{w}, z)_H d\rho_t(\omega) + \\ + \rho_t(\omega)(z, dw_t)_H - (c_t(\omega), z)\,dt + (c_t(\omega), z)_H dt,$$

whence it follows that $\rho_t(\omega)(\tilde{w}_t, z)_H$ is representable by a stochastic integral, i.e. it is a local martingale. It is easy to see that the characteristics of $\tilde{w}_t$ and w_t coincide. This completes the proof of the lemma.

It follows from formula (9) that the relation

$$d\varphi(x_t) = [a(\varphi, x_t) + (b(\varphi, x_t), \rho_t^{-1}(\omega)c_t(\omega))_H]\,dt + \\ + (b(\varphi, x_t), d\tilde{w}_t)_H.$$

is satisfied on the probability space $\{\Omega, \mathfrak{F}, \tilde{P}\}$. Thus, if x_t satisfies the equation

$$d\varphi(x_t) = \tilde{a}(\varphi, x_t)\,dt + (\tilde{b}(\varphi, x_t), d\tilde{w}_t)_H,$$

on the probability space $\{\Omega, \mathfrak{F}, \tilde{P}\}$, then

$$\begin{aligned} \tilde{b}(\varphi, x) &= b(\varphi, x), \\ \tilde{a}(\varphi, x_t) &= a(\varphi, x_t) + (b(\varphi, x_t), \rho_t^{-1}(\omega)c_t(\omega))_H. \end{aligned} \tag{10}$$

We shall assume that relation (10) is satisfied for some function $c_t(\omega)$. Then it is also satisfied for

$$\tilde{c}_t(\omega) = \mathsf{M}(\rho_t^{-1}(\omega)c_t(\omega)/x_t) = c_t(x_t),$$

i.e. for all t, $\varphi \in D$,

$$\tilde{a}(\varphi, x_t) = a(\varphi, x_t) + (b(\varphi, x_t), c_t(x_t))_H.$$

THEOREM 2. Suppose x_t is a solution of equation (1) and that there exists a bounded measurable function $c(x)$ with values in H such that

$$a_1(\varphi, x) - a(\varphi, x) = (b(\varphi, x), c(x))_H. \tag{11}$$

Then there exists a solution of the equation

$$d\varphi(x_t^{(1)}) = a_1(\varphi, x_t^{(1)})\,dt + (b(\varphi, x_t^{(1)}), dw_t), \quad \varphi \in D, \quad x_0^{(1)} = x, \tag{12}$$

for which the measure μ_T^1 corresponding to it on $C_{[0,T]}(X)$ will be absolutely continuous relative to the measure μ_T corresponding to the solution $x(t)$ of equation (1) on $C_{[0,T]}(X)$, and, in addition, almost everywhere with respect to the measure μ_T,

$$\frac{d\mu_T^{(1)}}{d\mu_T}(x(\cdot))=\exp\left\{\int_0^T (c(x_s),\ dw(s))_H-\frac{1}{2}\int_0^T |c(x_s)|_H^2\,ds\right\}.$$

If the solution (1) is weakly unique, then the solution (12) will also be weakly unique, and in this case formula (13) is valid for arbitrary solution (1) and (12).

Proof. Suppose $\rho_t(\omega)$ is given by the right member of (13) if we set $T = t$. Then

$$d\rho_t(\omega)=\rho_t(\omega)(c(x_s),\ dw(s))_H.$$

Moreover,

$$\mathsf{M}\rho_t^2(\omega)\leqslant \mathsf{M}\exp\left\{2\int_0^t (c(x_s),\ dw(s))_H\right\}$$

in virtue of the boundedness of $|c(x)|_H$ (see [15, 3, p.227]). Therefore, $\rho_t(\omega) = 1$ and we can assert that the measure $\mu_T^{(1)}$ for which (13) is satisfied corresponds to the solution of equation (12) if, in it, the process w_t is replaced by the process

$$\tilde{w}_t=w_t-\int_0^t c(x_s)\,ds,$$

which is a Wiener process on the probability space $\{\Omega, \mathfrak{S}, \mathsf{P}\}$, and

$$\tilde{\mathsf{P}}(A)=\int_A \rho_T(\omega)\,\mathsf{P}(d\omega).$$

By the same token, the existence of a weak solution is proven. In exactly the same way, if the measure $\bar{\mu}_T^{(1)}$ corresponds to some weak solution of (12), we find that the measure $\bar{\mu}_T$, for which

$$\frac{d\bar{\mu}_T}{d\bar{\mu}_T^{(1)}}(x^{(1)}(\cdot))=\exp\left\{-\int_0^T (c(x_s^{(1)}),\ dw(s))_H-\frac{1}{2}\int_0^T |c(x_s^{(1)})|_H^2\,ds\right\},$$

will correspond to some solution of (1). If μ_T^1 and $\bar{\mu}_T^1$ are distinct, then the measures μ_T and $\bar{\mu}_T$ will also be distinct. Thus, there follows from the uniqueness of the weak solution of (1) the weak uniqueness of the solution of (12). This completes the proof of the theorem.

REMARK. If equation (1) has a unique (strong) solution, then concerning equation (12) one can assert that it has only a weakly unique solution.

We shall now consider a transformation of the phase space whereby an equation of type (1) again goes over into an equation of the same type. Let $\tilde{X}$ be some other compactum and let S(x) be a continuous function that maps X onto $\tilde{X}$. We shall be interested in the conditions under which, for the process $\tilde{x}_t = S(x_t)$, one can write an equation of type (1) if x_t is a soltuion of (1). We denote by $\tilde{D}$ the set of functions φ from $C_{\tilde{X}}$ for which $\tilde{\varphi}(S(x)) \in D$. Then, for $\tilde{\varphi} \in \tilde{D}$, we shall have

$$d\tilde{\varphi}(\tilde{x}_t) = a(\tilde{\varphi}(S), x_t)\,dt + (b(\tilde{\varphi}(s), x_t), dw_t)_H. \tag{14}$$

THEOREM 3. If there exists a $\tilde{D}_0 \subset \tilde{D}$ such that, for $\tilde{\varphi} \in \tilde{D}_0$,

$$a(\tilde{\varphi}(S), x) = \tilde{a}(\tilde{\varphi}, S(x)), \quad b(\tilde{\varphi}(S), x) = U(\tilde{\varphi}, x)\,\tilde{b}(\tilde{\varphi}, S(x)),$$

where $\tilde{a}(\varphi, \tilde{x})$, $\tilde{b}(\varphi, \tilde{x})$ are linear with respect to $\tilde{\varphi}$ and measurable, take on values in R, H, resp., and $U(\tilde{\varphi}, \tilde{x})$ is in the space of unitary operators on H, then $\tilde{x}_t$ satisfies the relation

$$d\tilde{\varphi}(\tilde{x}_t) = \tilde{a}(\tilde{\varphi}, \tilde{x}_t)\,dt + (\tilde{b}(\tilde{\varphi}, \tilde{x}_t), d\tilde{w}_t), \; \tilde{\varphi} \in \tilde{D}_0, \quad \tilde{x}_0 = S(x), \tag{15}$$

where $\tilde{w}_t$ is some Wiener process.

The proof follows directly from (14) if one substitutes there $a(\tilde{\varphi}(s), x)$, $b(\tilde{\varphi}(s), x)$ and takes

$$\tilde{w}_t = \int_0^t U(\tilde{\varphi}, x_s)\,dw(s).$$

as the Wiener process $\tilde{w}_t$. But the fact that this process is a Wiener process in H follows from the fact that for $z \in H$

$$(z, \tilde{w}_t) = \int_0^t (U^{-1}(\tilde{\varphi}, x_s)\,z, dw_s)_H$$

is a martingale with the characteristic

$$\int_0^t |U^{-1}(\varphi, x_s)\,z|_H^2\,ds = (z, z)\,t.$$

This completes the proof of the theorem.

REMARK. If S maps X onto $\tilde{X}$ in on-to-one fashion, then one can take as $\tilde{D}$ the set of those $\tilde{\varphi}$ for which $\tilde{\varphi}(S^{-1}) \in D$, $\tilde{a}(\tilde{\varphi}, \tilde{x}) = a(\varphi(s), S^{-1}(x))$, $\tilde{b}(\tilde{\varphi}, \tilde{x}) = b(\tilde{\varphi}(S), S^{-1}(\tilde{x}))$, $U = J$, where J is the identity operator.

6. Stochastic Equations in R^m

In this section, we consider equations for processes on compacta in R^m. Basically, we shall be interested in processes that cannot be described with the aid of classical diffusion equations. A classical diffusion process is defined in the entire space R^m, the set D consists of twice continuously differentiable functions,

$$a(\varphi, x) = \sum_{i=1}^{m} a_i(x) \frac{\partial \varphi}{\partial x^i} + \frac{1}{2} \sum_{i,j=1}^{m} b_{ij}(x) \frac{\partial^2 \varphi}{\partial x^i \partial x^j},$$

$$b(\varphi, x) = \left(\sum_{i=1}^{m} \sigma_{i1}(x) \frac{\partial \varphi}{\partial x^i}, \ldots, \sum_{i=1}^{m} \sigma_{im}(x) \frac{\partial \varphi}{\partial x^i} \right),$$

where $a_i(x)$, $b_{ij}(x)$ are certain continuous functions on R^m, the matrix $(b_{ij}(x))_{i,j=1,\ldots,m}$ is nonnegative definite, the matrix $(\sigma_{ij}(x))_{i,=1,\ldots,m}$ is the nonnegative square root of the matrix $(b_{ij}(x))$, the vector of $b(\varphi, x)$ belongs to R^m (its coordinates are written in brackets), and in addition $H = R^m$. If $w_1(t), \ldots, w_m(t)$ are independent one-dimensional diffusion process, then the other is defined by the system of stochastic equations

$$dx^i(t) = a_i(x(t))\,dt + \sum_{j=1}^{m} \sigma_{ij}(x(t))\,dw_j(t); \tag{1}$$

where $x^i(t)$ is the i-th coordinate of the vector $x(t)$.

Very close to the classical case is the problem of investigating a stochastic differential equation on a smooth manifold (which can be considered as some subset of a finite-dimensional Euclidean space). Considering neighborhoods in which there exist coordinates, for these coordinates we can write a system of the form (1). Such a system is solved up to the moment of exit from this neighborhood, after which we go over to a new system of cordinates, and so on. We note that in this case it is already natural to consider the equation in the form that was considered in §§3 and 5, and in addition the set D must contain functions such that from these functions one could construct a system of coordinates for an arbitrary neighborhood.

Both cases considered are characterized by the fact that in D there enter smooth functions and the set X is locally Euclidean. The first peculiarity is disturbed for one-to-one continuous, but not smooth, mappings of the space. Such a transformation is a special case of the transformation of a stochastic differential equation considered in Theorem 3, §5. More interesting from a physical point of view are disturbances of local Euclideanness. We shall point out some cases that are possible here.

1. Occurrence of a boundary in the phase space. In this case, one can give the behavior of the process on the boundary: engulfing or various kinds of reflections. And the very presence of a boundary and the nature of the influence of the boundary on the subsequent motion have a completely definite physical meaning. We note that an engulfing boundary does not lead to new mathematical difficulties inasmuch as the local approach to the solution of an equation is itself based on the construction of a solution up to the moment of exit of the process from the region, i.e. its nonfalling on some boundary. Processes with reflection on the boundary under the presence of random perturbations differ essentially from deterministic processes with a boundary. The latter admit the integralwise solution of equations defined on them: we solve the equation until the solution arrives at the boundary, the influence of the boundary allows us to find boundary conditions for the reflection of a trajectory, which is then described with aid of the equation in the interior of the region, until the trajectory again exits to the boundary, and so on. In the stochastic case, the trajectory of the particle, egressing from the boundary, must, generally speaking, return to the boundary an infinite number of times before it departs from it to the extent that the influence of the boundary ceases to manifest itself. Therefore, here an intervalwise solution is not possible.

2. Appearance or disappearance in a system of one or several degrees of freedom. An example of such a motion is the diffusion of molecules which can dissociate to ions: upon the union of two ions, several degrees of freedom are lost in the molecule; upon the decomposition of molecules into ions, new degrees of freedom appear. In this example also, the motion itself and transitions of the system in a state with a different number of states the freedom occur in a random fashion. In classical mechanics, one considers systems for which there appear (or decompose) connections, i.e. the number of degrees of freedom changes. The definition of motion of such systems can not be carried out intervalwise, because of the many intervals of time in which the number of degrees of freedom of the system is unchanged. As in case 1, for stochastic systems this method is not admissible since such systems, situated in that point of a phase space whose neighborhoods contain components of different dimensions (i.e. with different degrees of freedom) will go over from component to component an infinite number of times during an arbitrarily small time before they go some distance away from the initial point and base themselves on one of the components. For systems of the indicated type, the phase space represents the union of several manifolds of various dimensions, having common points local Euclideanness is disturbed at these points.

3. Manifolds with branching points (or self-intersections). Here, local Euclideanness is disturbed in branching points. Systems whose state is described by some set of continuously varying parameters and a discrete parameter that takes on a finite (or countable) set of values have a phase space of this kind. Each branch of the manifold corresponds to a definite state of the discrete parameter, and the value of the discrete parameter may vary at branching points of the system. In stochastic systems, as well as in the preceding cases, transition from

branch to branch does not take place instantaneously; the system will go from one branch to another an infinite number of times before it recedes at some distance from the branching point. The motion of an electron in the neighborhood of the branching of a conductor can serve as the simplest example of such a motion.

Heat diffusion is placed upon the systematic motion of the electron induced by an electrical field; then before choosing the direction of motion at the branching point it will repeatedly go through the branching point from one branch of the conductor to another.

Considered further are stochastic differential equations in phase spaces of the three types listed above, with the assumption that in a neighborhood of a regular point (this is an interior point having a Euclidean neighborhood) the process is a diffusion process.

Process with reflection on the boundary. Let G be a connected region with a smooth boundary Γ in the space R^m, $X = G \cup \Gamma$. On Γ there is given a vector field $\nu(x)$, $\nu(x) \in R^m$, $|\nu(x)| = 1$. If $n(x)$ is the internal normal to Γ at the point x, then $(\nu(x), n(x)) > 0$. In G there is given a second-order differential operator

$$a(\varphi, x) = \sum_{i=1}^{m} a_i(x) \frac{\partial \varphi}{\partial x^i} + \frac{1}{2} \sum_{i,j=1}^{m} b_{ij}(x) \frac{\partial^2 \varphi}{\partial x^i \partial x^j}. \tag{2}$$

Our goal is to construct a process x_t in X which will be a diffusion process locally in the interior of G and will satisfy equation (1) (the $\sigma_{ij}(x)$ are defined in terms of $b_{ij}(x)$ as shown above), and falling on the boundary at the point x it reflects from that point in the direction $\nu(x)$. The latter means that for $x_0 = x \in \Gamma$, $x_t - x_0 \sim |x_t - x_0| \nu(x_0)$. In order to write down the stochastic equation for the process with reflection on the boundary (see, e.g., [10, §23]) we used the boundary processes - nondecreasing continuous processes - increasing only at those points t at which $x_t \in \Gamma$. The equations used in §3 allow us to exclude boundary processes from consideration.

We shall denote by D the set of twice continuously differentiable functions in the region G for which, for $x \in \Gamma$, $(\varphi'(x), \nu(x)) = 0$; here, $\varphi'(x)$ is the vector-derivative of the function (its gradient). We shall say that x_t is a process with reflection on the boundary Γ in the direction $\nu(x)$ satisfying equation (1) in the interior of G, if for all $\varphi \in D$, the relation

$$d\varphi(x_t) = a(\varphi, x_t)\, dt + \sum_{j=1}^{m} \sum_{i=1}^{m} \sigma_{ij}(x_t) \frac{\partial \varphi}{\partial x^i}(x_t)\, dw_j(t). \tag{3}$$

is satisfied. In a neighborhood of each point of the boundary, one can perform a smooth transformation such that the boundary goes over into a piece of a hyperplane, and so that $\nu(x)$ coincides with the normal to the hyperplane. Therefore, using the localization principle, it suffices to consider the case when G is the half-space $x^1 > 0$, Γ is the hyperplane $x^1 = 0$ (here, $x^1, \ldots, x^m$ are the coordinates of the point x), $\nu(x)$ is the normal to this hyperplane.

THEOREM 1. Suppose that in equation (3) the functions $a_i(x)$, $\sigma_{ij}(x)$ are defined for $x \in G \cup \Gamma = \{x : x^1 \geq 0\}$, uniformly continuously bounded and the matrix (σ_{ij}) is nongenerate, and D consists of twice continuously differentiable functions in G with bounded derivatives, for which

$$\frac{\partial\varphi}{\partial x^1}(0, x^2, \ldots, x^m) = 0.$$

Then there exists a weakly unique solution of equation (3), satisfying the following additional condition: the Lebesgue measure of those instants of time at which the process finds itself on the boundary equals o (such a process is called a <u>process with instantaneous reflection</u>).

<u>Proof</u>. In the space $\tilde{X} = R^m$ we consider the equation

$$d\tilde{\varphi}(\tilde{x}_t) = \tilde{a}(\tilde{\varphi}, x_t) + \sum_{j=1}^{m}\sum_{i=1}^{m} \tilde{\sigma}_{ij}(\tilde{x}_t)\frac{\partial\tilde{\varphi}}{\partial x^i}(\tilde{x}_t)\, d\tilde{w}_j(t),$$

where

$$\tilde{a}(\tilde{\varphi}, x) = \sum_{i=1}^{m} \tilde{a}_i(x)\frac{\partial\tilde{\varphi}}{\partial x^i} + \frac{1}{2}\sum_{i,j,k=1}^{m} \tilde{\sigma}_{ik}(x)\sigma_{jk}(x)\frac{\partial^2\tilde{\varphi}}{\partial x^i \partial x^j},$$

$$\tilde{a}_i(x) = \tilde{a}_i(x^1, x^2, \ldots, x^m) = a_i(|x^1|, x^2, \ldots, x^m), \quad i > 1,$$
$$\tilde{a}_1(x) = \tilde{a}_1(x^1, x^2, \ldots, x^m) = a_1(|x^1|, x^2, \ldots, x^m)\operatorname{sign} x^1,$$
$$\tilde{\sigma}_{11}(x) = \tilde{\sigma}_{11}(x^1, x^2, \ldots, x^m) = \sigma_{11}(|x^1|, x^2, \ldots, x^m),$$
$$\tilde{\sigma}_{1i}(x) = \sigma_{1i}(|x^1|, x^2, \ldots, x^m)\operatorname{sign} x^1,$$
$$\tilde{\sigma}_{i1}(x) = \sigma_{i1}(|x^1|, x^2, \ldots, x^m)\operatorname{sign} x^1,$$
$$\tilde{\sigma}_{ij}(x) = \sigma_{ij}(|x^1|, x^2, \ldots, x^m), \quad i, j > 1,$$

$\tilde{D}$ is the set of twice continuously differentiable functions. The existence and uniqueness of the solution of equation (4) ($\tilde{a}$, generally speaking, is jumplike) follows from the results of Strook and Varadhan [35]. We shall denote by $\tilde{D}_o$ the set of those functions in $\tilde{D}$ which are even in x^1: (then $\frac{\partial\varphi}{\partial x^1}(0, x^2, \ldots, x^m) = 0$. Suppose $S(x)$ maps $\tilde{X}$ into $X = G \cup \Gamma$ by the formula $S(x) = (|x^1|, x^2, \ldots, x^m)$. For all $\varphi \in \tilde{D}_o \tilde{\varphi}(x) = \varphi(S(x))$, where $\varphi \in D$, and for $\varphi \in D$, $\varphi(S(x)) \in \tilde{D}_o$. Furthermore, inasmuch as the function $\frac{\partial f}{\partial x^1}$ is odd in f, even in x^1, and $\frac{\partial^2 f}{(\partial x^1)^2}$ is e even, then

$$\tilde{b}(\varphi(S), x) = \left(\sum_{i=1}^{m} \tilde{\sigma}_{i1}(x)\frac{\partial\varphi(S(x))}{\partial x^i}, \ldots, \sum_{i=1}^{m} \tilde{\sigma}_{im}(x)\frac{\partial\varphi(S(x))}{\partial x^i}\right) =$$

$$\left(\operatorname{sign} x^1\left[\sigma_{11}(S(x))\frac{\partial\varphi(S(x))}{\partial x^1}\operatorname{sign} x^1 + \sum_{i=2}^{m}\sigma_{i1}(S(x))\frac{\partial\varphi(S(x))}{\partial x^i}\right], \ldots\right.$$

$$\left.\ldots, \left[\sigma_{1m}(S(x))\frac{\partial\varphi(S(x))}{\partial x^1}\operatorname{sign} x^1 + \sum_{i=2}^{m}\sigma_{im}(S(x))\frac{\partial\varphi(S(x))}{\partial x^i}\right]\right) =$$

$$= U(x)\,b(\varphi, S(x)),$$

where the unitary operator U(x) in the natural basis of R^m has a matrix with elements $u_{ij}(x) = \delta_{ij}$ for $i + j > 2$, $u_{11}(x) = \operatorname{sign} x^1$. If we set $w_k(t) = \tilde{w}_k(t)$ for $k = 2, 3, \ldots, m$, $w_1(t) = \int_0^t \operatorname{sign} \tilde{x}_s^1 \, d\tilde{w}_1(s)$, then, by virtue of Theorem 3, §5, the process $x_t = S(\tilde{x}_t)$ satisfies equation (3). The existence and weak uniqueness of the solution of equation (4) imply the existence and weak uniqueness of the solution of equation (3). To this end, it suffices to prove that every solution of equation (3) is representable in the form $S(\tilde{x}_t)$, where $\tilde{x}_t$ is the solution of equation (4). In fact, let x_t be the solution of equation (3). We introduce the random function δ_t possessing the following properties: a) δ_t takes on the values ±1 and 0, whereby $\delta_t = 0$ if and only if $x_t \in \Gamma$, and δ_t is continuous at all points where $\delta_t \neq 0$; b) the conditional distributions δ_s, $s \leq t$ and $-\delta_s$, $s \leq t$ for given x_s, $s \leq t$ are the same. We set $\tilde{x}_t = (\delta_t x_t^1, x_t^2, \ldots, x_t^m)$. Then for the twice differentiable function $\tilde{\varphi}(x)$, defined on R^m, for which $\frac{\partial\tilde{\varphi}}{\partial x^1}(0, x^2, \ldots, x^m) = 0$, we can write at all points t for which $x_t \bar{\in} \Gamma$:

$$d\tilde{\varphi}(\tilde{x}_t) = \chi_{\{\delta_t=+1\}}\, d\varphi_1(x_t) + \chi_{\{\delta_t=-1\}}\, d\varphi_2(x_t),$$

where $\varphi_1(x) = \tilde{\varphi}(x)$, $x \in X$, $\varphi_2(x) = \tilde{\varphi}(-x^1 \cdot x^2, \ldots, x^m)$, $x \in X$, $\varphi_1, \varphi_2 \in D$. Therefore, by virtue of (3):

$$d\tilde{\varphi}(\tilde{x}_t) = \chi_{\{\delta_t=+1\}}\, a(\varphi_1, \tilde{x}_t)\,dt + \chi_{\{\delta_t=-1\}}\, a(\varphi_2, \tilde{x}_t)\,dt + \sum_{j=1}^{m}\sum_{i=1}^{m}\Big[\sigma_{ij}(\tilde{x}_t)\frac{\partial\varphi_1}{\partial x^i}(\tilde{x}_t)\chi_{\{\delta_t=+1\}} + \sigma_{ij}(S(\tilde{x}_t))\frac{\partial\varphi_2}{\partial x^i}(S(\tilde{x}_t))\chi_{\{\delta_t=-1\}}\Big]dw_j(t).$$

Inasmuch as $\delta_t = \operatorname{sign} x_t^1$, then

$$\chi_{\{\delta_t=+1\}}\, a(\varphi_1, S(\tilde{x}_t)) + \chi_{\{\delta_t=-1\}}\, a(\varphi_2, S(\tilde{x}_t)) = \tilde{a}(\tilde{\varphi}, \tilde{x}_t),$$

$$\sum_{i=1}^{m}\Big[\sigma_{ij}(\tilde{x}_t)\frac{\partial\varphi_1}{\partial x^i}(\tilde{x}_t)\chi_{\{\delta_t=+1\}} + \sigma_{ij}(S(\tilde{x}_t))\frac{\partial\varphi_2}{\partial x^i}(S(\tilde{x}_t))\chi_{\{\delta_t=-1\}}\Big] = \begin{cases}\displaystyle\sum_{i=1}^{m}\tilde{\sigma}_{ij}(\tilde{x}_t)\frac{\partial\tilde{\varphi}}{\partial x^i}(\tilde{x}_t), & j>1;\\ \displaystyle\operatorname{sign}\tilde{x}_t^1\sum_{i=1}^{m}\tilde{\sigma}_{i1}(\tilde{x}_t)\frac{\partial\tilde{\varphi}}{\partial x^i}(\tilde{x}_t), & j=1.\end{cases}$$

Consequently, for all t for which $\delta_t \neq 0$ equation (4) is satisfied. It remains to note that the Lebesgue measure of those t for which $\delta_t = 0$ equals zero. Therefore, equation (4) can be integrated, i.e. it is valid in general in virtue of the definition of a stochastic differential. This completes the proof of the theorem.

Processes with a varying number of degrees of freedom. We consider a point a neighborhood of which contains components of the phase space of different dimensions. As will be shown below, it suffices to consider the junction of two components - they can be joined to one. It may turn out that the juncture point is situated on the boundary of one (or both) components. Such processes can be obtained by the method discussed in Theorem 1, from processes with a boundary. We shall introduce characteristic examples of the juncture of two-dimensional and one-dimensional components:

(1) $X = \{(x^1, x^2) : x^1 \geqq 0\} \cup \{(x^1, x^2) : x^2 = 0\}$

here both components have a boundary;

(2) $X = \{(x^1, x^2, x^3) : x^3 = 0\} \cup \{(x^1, x^2, x^3) : x^1 = 0, x^2 = 0, x^3 > 0\}$

the one-dimensional component has a boundary;

(3) $X = \{(x^1, x^2, x^3= : x^3 = 0, x^1 \geq 0\} \cup \{(x^1, x^2, x^3) : x^1 = 0, x^2 = 0\}$

the two-dimensional component has a boundary;

(4) $X = \{(x^1, x^2, x^3) : x^3 = 0\} \cup \{(x^1, x^2, x^3) : x^1 = 0, x^3 = 0\}$

a boundary is absent.

From a process of the form (4) one can obtain processes of the remaining forms, replacing x^1 by $|x^1|$, or x^3 by $|x^3|$. In this connection, one must select the coefficients in the region (4) so that the indicated mapping allows application of Theorem 3, §5. It was pointed out in Theorem 1 how to choose coefficients upon replacing x^1 by $|x^1|$.

We shall consider the stochastic equation in region (4). Suppose H consists of functions of the set D, given on X and twice continuously differentiable on each component. As long as the process does not fall into the juncture point, then it is an ordinary diffusion process on each component. In this case, one can take R^2 for H. We set

$$a(\varphi, x) = \begin{cases} a_1(x)\dfrac{\partial\varphi}{\partial x^1} + a_2(x)\dfrac{\partial\varphi}{\partial x^2} + \dfrac{1}{2}\Big[b_{11}(x)\dfrac{\partial^2\varphi}{(\partial x^1)^2} + \\ + 2b_{12}(x)\dfrac{\partial^2\varphi}{\partial x^1\partial x^2} + b_{22}(x)\dfrac{\partial^2\varphi}{(\partial x^2)^2}\Big], \quad x_3 = 0; \\ a_3(x)\dfrac{\partial\varphi}{\partial x^3} + \dfrac{1}{2}b_{33}(x)\dfrac{\partial^2\varphi}{\partial x^3}, \quad x_1 = x_2 = 0; \end{cases}$$

$$b(\varphi, x) = \begin{cases} \Big(\sigma_{11}(x)\dfrac{\partial\varphi}{\partial x^1} + \sigma_{12}(x)\dfrac{\partial\varphi}{\partial x^2}, \\ \sigma_{21}(x)\dfrac{\partial\varphi}{\partial x^1} + \sigma_{22}\dfrac{\partial\varphi}{\partial x^2}\Big), \quad x_3 = 0; \\ \Big(\sigma_{23}(x)\dfrac{\partial\varphi}{\partial x^3}, \quad \sigma_{23}(x)\dfrac{\partial\varphi}{\partial x^3}\Big), \quad x_1 = x_2 = 0; \end{cases} \tag{5}$$

$$\begin{pmatrix}\sigma_{11} & \sigma_{12} \\ \sigma_{21} & \sigma_{22}\end{pmatrix}^2 = \begin{pmatrix} b_{11} & b_{12} \\ b_{12} & b_{22}\end{pmatrix}, \qquad \sigma_{13}^2 = \sigma_{23}^2 = b_{33}.$$

Then on each component the solution of the equation

$$d\varphi(x_t) = a(\varphi, x_t)\,dt + (b(\varphi, x_t), dw(t)), \tag{6}$$

where $w(t) = (w_1(t), w_2(t))$ is a Wiener process in R^2, will be a diffusion process. Now we must only determine the behavior of the process at the juncture point, i.e. the nature of the transition of the process from component to component. The nature of the transition is determined by the behavior of the function in D at the juncture point, which is expressed by means of definite conditions of conjugation of the derivations of functions φ on different components at the juncture point. To explain the possible conjugation conditions, we shall show how it is possible to obtain equation (6) with coefficients of the form (5) with the aid of mapping the spaces R^4 and X and Theorem 3, §5. Let S map R^4 into X by the formula

$$S(x^1, x^2, x^3, x^4) = (x^1, x^2, 0) \text{ for } x^4 > 0,$$

$$S(x^1, x^2, x^3, x^4) = (0, 0, x^3) \text{ for } x^4 < 0.$$

Then, for $\varphi \in D$, the function $\tilde{\varphi}(x^1, x^2, x^3, x^4) = \varphi(S(x^1, x^2, x^3, x^4))$ will be twice continuously differentiable for $x^4 \neq 0$. If there is a process $\tilde{x}_t$ in R^4 such that the process $x_t = S(\tilde{x}_t)$ satisfies equation (6), then the coefficients $\tilde{a}(\tilde{\varphi}, x)$, $\tilde{b}(\tilde{\varphi}, x)$ of the equation for $\tilde{x}_t$ must satisfy the relations

$$\tilde{a}(\varphi(S), x^1, x^2, x^3, x^4) = a(\varphi, S(x^1, x^2, x^3, x^4)),$$
$$\tilde{b}(\varphi(S), x^1, x^2, x^3, x^4) = U(x)\, b(\varphi, S(x^1, x^2, x^3, x^4)),$$

where $U(x^1, x^2, x^3, x^4)$ is a unitary operator in R^2. Moreover, the process $\tilde{x}_t$ must possess the following property: for $\tilde{x}_t^i > 0$ and $(\tilde{x}_t^1)^2 + (\tilde{x}_t^2)^2 \to 0$ then we must also have $\tilde{x}_t^3 \to 0$, $\tilde{x}_t^1 \to 0$; and for $\tilde{x}_t^4 < 0$ and $|x_t^3| \to 0$ we must also have $(\tilde{x}_t^1)^2 + (\tilde{x}_t^2)^2 \to 0$ and $\tilde{x}_t^4 \to 0$. These requirements follow from the continuity condition for the processes $\tilde{x}_t$ and $S(\tilde{x}_t)$ for $\tilde{x}_t^4 = 0$. Let

$$\tilde{a}_i(x^1, x^2, x^3, x^4) = \begin{cases} a_i(x^1, x^2), & x^4 > 0; \\ 0, & x^4 < 0, \quad i = 1, 2; \end{cases}$$

$$\tilde{a}_3(x^1, x^2, x^3, x^4) = \begin{cases} 0, & x^4 > 0; \\ a_3(x^3), & x^4 < 0; \end{cases}$$

$$\tilde{a}_4(x^1, x^2, x^3, x^4) = \begin{cases} x^1 a_1(x^1, x^2) + x^2 a_2(x^1, x^2) + \\ \quad + \sum_{i,k=1}^{2} \sigma_{ik}^2(x^1, x^2), & x^4 > 0; \\ -x^3 a_3(x^3) - (\sigma_{13}^2(x^3) + \sigma_{23}^2(x^3)), & x^4 < 0; \end{cases}$$

$$\tilde{\sigma}_{ik}(x^1, x^2, x^3, x^4) = \begin{cases} \sigma_{ik}(x^1, x^2), & x^4 > 0; \\ 0, & x^4 < 0; \end{cases} \quad i = 1, 2;\ k = 1, 2;$$

$$\tilde{\sigma}_{i3}(x^1, x^2, x^3, x^4) = \begin{cases} 0, & x^4 > 0; \\ \sigma_{i3}(x^3), & x^4 < 0; \end{cases} \quad i = 1, 2;$$

$$\tilde{\sigma}_{i4}(x^1, x^2, x^3, x^4) = \begin{cases} x^1 \sigma_{i1}(x^1, x^2) + x^2 \sigma_{i2}(x^1, x^2), & x^4 > 0; \\ -x^3 \sigma_{i3}(x^3), & x^4 < 0. \end{cases}$$

We shall consider the system of stochastic differential equations

$$d\tilde{x}_t^k = \tilde{a}_k(\tilde{x}_t) + \tilde{\sigma}_{1k}(\tilde{x}_t)\, d\tilde{w}_1(t) + \tilde{\sigma}_{2k}(\tilde{x}_t)\, d\tilde{w}_k(t), \tag{8}$$

which can be written in the form

$$d\tilde{\varphi}(\tilde{x}_t) = \tilde{a}(\tilde{\varphi}, \tilde{x}_t)\, dt + (\tilde{b}(\tilde{\varphi}, \tilde{x}_t), d\tilde{w}_t), \tag{9}$$

where

$$\tilde{a}(\tilde{\varphi}, x) = \sum_{k=1}^{4} \tilde{a}_k(x) \frac{\partial \tilde{\varphi}}{\partial x^k} + \frac{1}{2} \sum_{i,j=1}^{4} \tilde{b}_{ij}(x) \frac{\partial^2 \varphi}{\partial x^i \partial x^j},$$

$$\tilde{b}(\varphi, x) = \left(\sum_{k=1}^{4} \tilde{\sigma}_{1k}(x) \frac{\partial \tilde{\varphi}}{\partial x^k}, \ \sum_{k=1}^{4} \tilde{\sigma}_{2k}(x) \frac{\partial \tilde{\varphi}}{\partial x^k} \right),$$

$$\tilde{b}_{ij}(x) = \sum_{k=1}^{2} \sigma_{ki}(x) \sigma_{kj}(x).$$

If x_t is the solution of equation (8), then, using Itô's formula and the form of the coefficients of the equation, we can verify that

$$d[(\tilde{x}_t^1)^2 + (\tilde{x}_t^2)^2 - \tilde{x}_t^4] = 0 \quad \text{for } \tilde{x}_t^4 > 0,$$

$$d[(\tilde{x}_t^3)^2 + \tilde{x}_t^1] = 0 \quad \text{for } \tilde{x}_t^4 < 0.$$

Suppose the initial conditions are such that $(\tilde{x}_0^1)^2 + (\tilde{x}_0^2)^2 - \tilde{x}_0^4 = 0$ for $\tilde{x}_0^4 > 0$, or that $(\tilde{x}_0^3) = -\tilde{x}_0^4$ for $\tilde{x}_0^4 < 0$. Then

$$(\tilde{x}_t^1)^2 + (\tilde{x}_t^2)^2 - \chi_{\{\tilde{x}_t^4 > 0\}} \tilde{x}_t^4 = 0,$$

$$(\tilde{x}_t^3)^2 + \chi_{\{\tilde{x}_t^4 < 0\}} \tilde{x}_t^4 = 0.$$

Therefore, the process $S(\tilde{x}_t) = x_t$ is continuous. Relations (7) are satisfied for it if U is the identity operator. On the basis of Theorem 3, §5, the process x_t will be a solution of equation (6) with the coefficients (5). Equation (8) has a singularity for $x^4 = 0$ inasmuch as in this case $\tilde{\sigma}_{i4} = 0$ with probability 1 (for $\tilde{x}_t^4 = 0$ and $(\tilde{x}_t^1)^2 + (\tilde{x}_t^2)^2 + (\tilde{x}_t^3) = 0$), $\tilde{a}_4$ changes sign. Therefore, the solution of equation (8) for given initial condition, generally speaking, is not unique. Describing the possible solutions of this equation with the aid of the constructed mapping one can obtain all solutions of equation (6) inasmuch as the mapping S maps on one-to-one fashion the range of the process $\bar{x}_t$ on X: $S^{-1}(x) = (x^1, x^2, 0, (x^1)^2 + (x^2)^2)$ for $x^3 = 0$, $S^{-1}(x) = (0, 0, x^3, -(x^3)^2)$ for $x_1 = 0$, $x_2 = 0$. We note, first of all, that in this case, when the point (0, 0) on the (x_1, x_2)-plane is not accessible, then the process transpires on one component and the solution of equation (8) is unique. But if the point O is accessible, then in a neighborhood of the point O one can define a positive harmonic function everywhere except at the point O, equal to zero at the point O (this will be, for example, the probability of attaining the boundary of the unit circle sooner than the point O). We shall denote it by $\varphi_1(x)$. Moreover, the point O must be such that the process could start at the point O. This signifies that the point O is regular. If it is also regular for a one-dimensional process on the straight line $(0, 0, x^3)$, then there exists a positive function $\varphi_2(x^3)$ is harmonic everywhere except at the point O. Let x_t be the solution of equation (6), being a Markov process. If ψ is a harmonic function for this process in a neighborhood of the point O and $\psi > 0$ on $(0, 0, x^3)$, $\psi < 0$ on $(x^1, x^2, 0)$, then, for some $c_1 > 0$ and $c_2 > 0$, $\psi(x^1, x^2, 0) = -c_1\varphi_1(x^1, x^2)$, $\psi(0, 0, x^3) = c_2\varphi_2(x^3)$. Now as D one can choose the set of functions φ such that, besides satisfying the above-mentioned smoothness conditions, can also satisfy the following condition: for each φ there exists a c such that $\varphi - c\psi$ has first derivatives equal to zero at the point O.

A detailed investigation of the questions of existence and uniqueness for equations of the type (6) or (8) must be based on a developed theory of the behavior of solutions of stochastic equations in neighborhoods of a singular point. However, here we have studied completely only the one-dimensional case.

We shall now consider the juncture of two spaces of different dimension in the general case. Let $X = (R^n \times R^m) \cup (R^1 \times R^m)$, whereby the points $(0, z)$, $0 \in R^n$, $z \in R^m$ and $(0, z)$, $0 \in R^1$, $z \in R^m$ are identified. In other words, X is considered as a subset of the space $R^n \times R^m \times R^1$, consisting of points of the form (x, z, y), $x \in R^m$, $z \in R^m$, $y \in R^1$, for which either $x = 0$ or $y = 0$. If L_1 consists of the points $(x, z, 0)$ and L_2 of the points $(0, z, y)$, then stochastic differential equations of the form we considered above are defined on L_1 and L_2. By the same token, an equation is defined on $L_1 \cup L_2$: if φ is a twice continuously differentiable function on each component, then for $(x_t, z_t, y_t) \bar{\in} L_1 \cap L_2$, one can write down an equation for this process using equations on the individual components. In order to obtain this equation with the aid of transformations of the space from equations in Euclidean space, we introduce one more coordinate $u \in R$:

$$u_t = |x_t|^2 \text{ if } y_t = 0; \quad u_t = -|y_t|^2 \text{ if } x_t = 0,$$

where (x_t, z_t, y_t) is a solution of the initial equation. Now the equation can be rewritten in $R^n \times R^m \times R^1 \times R$ in the form of an ordinary equation and, with the aid of mapping

$$S(x_t, z_t, y_t, u_t) = \begin{cases} (x_t, z_t, 0), & u_t > 0; \\ (0, z_t, y_t), & u_t < 0 \end{cases}$$

carry it over into the initial equation. If (x_t, z_t, y_t, u_t) is the solution of the equation with initial condition $u_o = |x_o|^2$, $y_o = 0$ or $u_o = -|y_o|^2$, $x_o = 0$, the the mapping S carried this solution over into a continuous process, which will satisfy the initial equation. We note for $u = 0$ we obtain a same kind of singularity as above.

Branching of the phase space. This rather simply reduces to the consideration of the junctures of spaces of different dimensions and takes into consideration the influence of the boundary. We shall demonstrate this by means of a simple example - the juncture of three one-dimensional branches: $X = \{1, 2, 3\} \times (0, \infty) \cup \{0\}$ - this is the union of three rays, the point O belongs to each. Let $\tilde{X}$ be the set in three-dimensional space consisting of the plane $(x^1, x^2, 0)$ and the straight line $(x^1)^2 + (x^2)^2 = 0$. The mapping S of $\tilde{X}$ into X defined by the equations $S(0, 0, 0) = 0$, $S(x^1, x^2, 0) = (1(x^1)^2 + (x^2)^2$ for $(x^1)^2 + (x^2)^2 > 0$, $S(0, 0, x^3) = (2, x^3)$ for $x^3 > 0$, $S(0, 0, x^3) = (3, -x^3)$ for $x^3 < 0$ is continuous, if X is equipped with the natural topology. Therefore, the equation for the process in X can be obtained with the aid of a mapping of the equation in $\tilde{X}$; we have already considered an equation of the latter type.

Chapter 2

RANDOMLY INTERACTING SYSTEMS OF PARTICLES

In this chapter we consider systems consisting of a large number of like particles that are moving according to the laws of mechanics under the influences of an external force field and also of forces of mutual interaction between particles. The interaction forces between particles have two components: one is remote action (for example, forces of gravity, electric and electromagnetic forces), which is assumed to be nonrandom; the other is near action (the result of collision), which is of the nature of an impulse, the velocity of particles under such an interaction varies jumpwise and is assumed to be random. The random nature of the latter interaction can be explained by the complexity of the geometric form of the particles, the presence of quantum mechanical effects, the presence of internal degrees of freedom to which a part of collision energy may go, and so forth. Our goal is to investigate the limiting behavior of such systems under the assumption that the number of particles increases without limit, and their total mass remains invariant, in which connection the dimensions of the particles vary correspondingly. It turns out that there exists a limit distribution for the mass of particles with respect to volume and in the study of the motion of an individual particle it suffices to know only this distribution and not the motion of each particle; that is, the action of the remaining particles on the given one is averaged. If we then assume that the quantity of mutually interacting particles on the average in a unit of time increases to infinity, then in the limit we obtain the Brownian motion process. Thus, the situation that for a diffusion process we have a first-order equation, and a second-order Newton equation for the description of the motion of particles finds a clarification.

1. Stochastic Equation for Systems of Randomly Interacting Particles

Let X be a phase space in which a motion of particles takes place. This space may be simply three-dimensional if, for the description of the motion, it suffices to describe the motion of the center of gravity or a more complicated space if the particles have some spatial form (if they are considered as solid bodies, then besides the coordinates of the center of gravity one can furthermore have Euler angles; if the form of the body varies with time, then one must furthermore prescribe parameters that determine the form of the body). We denote by V the velocity space. For linear X, we can assume that the space V coincides with X; for more complicated X they are, in general, distinct. However, if X can be embedded in some linear space, then V will also lie in the same linear space. Therefore, in the sequel we shall assume that X is a linear space and that V coincides with X, and for the velocity space we shall use the same symbol, X.

Suppose there are n particles in X. We shall denote the position of the i-th particle by x_i (by $x_i(t)$ at the moment t) and its velocity by v_i (by $v_i(t)$ at the moment t). Let A(x, v) be the force of the external field acting on a particle occurring in the position x and having velocity v, and let a(x, y, v, u) be the force of the 'remote action', interacting with the same particle from the side of the particle occurring at the position y and having velocity u. To describe the 'near action' we shall assume that the interaction of distinct pairs of particles occurs independently and we shall use n(n-1)/2 (the number of pairs) independent Poisson measures which will thus define the magnitude and moment of time of the random interaction. In general, to characterize a random interaction, one must give the conditional probability of the situation that during time Δt the first particle changes its velocity by a given magnitude Δv under the condition that the positions and velocities of both particles are given. If, for Borel sets $C \subset X$,

$$\mathsf{P}\{\Delta v \in C/x, y, u, v\} = \mu(C, x, y, u, v)\,\Delta t, \tag{1}$$

where μ is a finite measure on X, then the conditional distribution coincides to within o(Δt) with the distribution of the random variable

$$\int f(\theta, x, y, u, v)\, p(d\theta \times dt), \tag{2}$$

where $p(d\theta \times dt)$ is a Poisson random measure with independent values, defined on the σ-algebra $\mathfrak{C} \times \mathfrak{A}_+$ of the space $\Theta \times R_+$, $(\Theta, \mathfrak{C})$ is some measurable space; the function of (θ, x, y, u, v) is a measurable function on $\Theta \times X^4$ with values in X. Concerning the Poisson measure, we assume that

$$\mathsf{M}p(d\theta \times dt) = m(d\theta)\,dt,$$

where m is some measure on Θ and f is such that

$$m(\{\theta: \quad f(\theta, x, y, v, u) \in C\}) = \mu(C, x, y, u, v) \tag{3}$$

(the existence of such a function f for an arbitrary measure μ and the nature of the dependence of f on the parameters was discussed in §2, CHapter 1). Inasmuch as for sufficiently small Δt to within o(Δt) the integral (2) equals zero with probability 1-m(Θ)Δt, and it coincides with $f(\tilde{\theta}, x, y, v, u)$ with probability m(Θ)Δt, where $\tilde{\theta}$ is a random quantity in Θ for which

$$P\{\tilde{\theta} \in C\} = \frac{m(C)}{m(\theta)},$$

then it follows from (3) that the distribution (2) coincides with the right member of (1). We denote by $P_{ij}(d\theta \times dt)$ the Poisson measure that defines the interaction of the i-th and j-th particles, $P_{ij} = P_{ji}$; for distinct pairs (i, j), these measures are independent. Thus, if the interaction of the i-th and j-th particles occurred on the interval [t, t + dt], then $v_i(t)$ has undergone the jump

$$\int_\Theta f(\theta,\ x_i(t),\ x_j(t),\ v_i(t),\ v_j(t))\, p_{ij}(d\theta \times dt).$$

It follows from the law of conservation of the quantity of motion that $f(\theta, x, y, v, u) = -f(\theta, y, x, u, v)$. We can now write the system of stochastic differential equations for the system of particles:

$$dv_i(t) = \Big[A(x_i,\ v_i) + \sum_j a(x_i,\ x_j,\ v_i,\ v_j)\Big] dt + $$
$$+ \sum_j \int_\Theta f(\theta,\ x_i,\ v_i,\ x_j,\ v_j)\, p_{ij}(d\theta \times dt),$$
$$dx_i(t) = v_i(t)\, dt$$

(the functions x_i, x_j, v_i, v_j occurring in the right member of (4) are also considered at the moment of time t, as is the case for the left member; under these conditions, when they occur in relations at various moments of time, we shall indicate the argument). Equations of type (4) were considered, for example, in [10] and [31]. The conditions for the existence and uniqueness of a solution of the system of equations (4) will, for instance, be the following

$$\sup_{x,y,v,u} \left\{ \frac{|A(x,\ v) - A(\bar{x},\ \bar{v})| + |a(x,\ y,\ v,\ u) - a(\bar{x},\ \bar{y},\ \bar{v},\ \bar{u})|}{|x - \bar{x}| + |v - \bar{v}| + |y - \bar{y}| + |u - \bar{u}|} + \right.$$
$$\left. + |f(\theta,\ x,\ v,\ y,\ u)| \right\} < \infty$$

and, moreover, $f(\theta, x, v, y, u)$ is continuous with respect to x, v, y, u with respect to the measure $m(d\theta)$.

We shall be interested in the 'statistical distribution function' of a system of particles defining a number of particles whose coordinates and velocities lie in a prescribed set at a given moment of time t. This is a measure $\mu_t(dx \times dv)$ given on X^2 by the following relation: for each continuous bounded function $g(x, v)$ on X^2,

$$\int g(x,\ v)\, \mu_t(dx \times dv) = \frac{1}{n} \sum_i g(x_i(t),\ v_i(t)). \tag{5}$$

It is more convenient to characterize the state of the particle by the pair (x; v) and to take $Z = X^2$ as the phase space. Let $\tilde{A}$, $\tilde{a}$, $\tilde{f}$ be functions with values in Z for which

$$\tilde{A}(z)=(v;\ A(x,\ v)),\quad \tilde{a}(z,\ z')=(0;\ a(x,\ v,\ x',\ v')),$$
$$\tilde{f}(\theta,\ z,\ z')=(0;\ f(\theta,\ x,\ v,\ x',\ v')),$$

where z = (x, v), z' = (x', v'). Then the system of equations (4) can be rewritten as

$$dz_i(t)=\Big[\tilde{A}(z_i)+\sum_j \tilde{a}(z_i,\ z_j)\Big]dt+ \\ +\sum_j \int_\Theta \tilde{f}(\theta,\ z_i,\ z_j)\,p_{ij}(d\theta\times dt). \tag{6}$$

Let $\varphi(z)$ be a continuously differentiable function that together with its derivative is bounded. Using the Itô formula, we can write

$$d\varphi(z_i(t))=\Big(\varphi'(z_i),\ \tilde{A}(z_i)+\sum_j \tilde{a}(z_i,\ z_j)\Big)dt+ \\ +\sum_j \int_\Theta [\varphi(z_i+\tilde{f}(\theta,\ z_i,\ z_j))-\varphi(z_i)]\,p_{ij}(d\theta\times dt). \tag{7}$$

We introduce the centered measures

$$q_{ij}(d\theta\times dt)=p_{ij}(d\theta\times dt)-m(d\theta)\,dt.$$

Expressing p_{ij} in terms of q_{ij} in equation (7) and then adding them with respect to i, we obtain the following relation for the statistical distribution function

$$d\int \varphi(z)\,\mu_t(dz)= \\ =\Big\{\int(\varphi'(z),\ \tilde{A}(z))\,\mu_t(dz)+n\int\int\Big\{(\varphi'(z),\ \tilde{a}(z,\ z'))+ \\ +\int_\Theta [\varphi(z+\tilde{f}(\theta,\ z,\ z'))-\varphi(z)]\,m(d\theta)\Big\}\mu_t(dz)\,\mu_t(dz')\Big\}dt+ \\ +\frac{1}{n}\sum_{i,j}\int[\varphi(z_i+\tilde{f}(\theta,\ z_i,\ z_j))-\varphi(z_i)]\,q_{ij}(d\theta\times dt) \tag{8}$$

(here, for convenience in writing, we assume that $\tilde{a}(z,\ z)=0$, $\tilde{f}(\theta,\ z,\ z)=0$; hence we assume that terms in the sum with i = j are equal to zero). The sum of integrals with respect to q_{ij} is the sum of martingales. We calculate the characteristic of the martingale

$$\zeta_n(\varphi,\ t)=\frac{1}{n}\sum_{i,j}\int_0^t\int_\Theta [\varphi(z_i(s)+ \\ +\tilde{f}(\theta,\ z_i(s),\ z_j(s)))-\varphi(z_i(s))]\,q_{ij}(d\theta\times ds). \tag{9}$$

Since $q_{ij} = q_{ji}$ and they are independent for distinct pairs and $\tilde{f}(\theta, z_i, z_j) = -\tilde{f}(\theta, z_j, z_i)$, then

$$\begin{aligned} &M\zeta_n^2(\varphi, t) = \\ &= M\frac{1}{n^2}\left(\sum_{i<j}\int_0^t\int_\Theta [\varphi(z_i(s) + \tilde{f}(\theta, z_i(s), z_j(s))) + \varphi(z_j(s) - \right. \\ &\left. - \tilde{f}(\theta, z_i(s), z_j(s))) - \varphi(z_i(s)) - \varphi(z_j(s))]\, q_{ij}(d\theta \times ds)\right)^2 = \\ &= \frac{1}{n^2}\sum_{i<j} M\int_0^t\int_\Theta [\varphi(z_i(s) + \tilde{f}(\theta, z_i(s), z_j(s))) + \varphi(z_j(s) - \\ &- \tilde{f}(\theta, z_i(s), z_j(s))) - \varphi(z_i(s)) - \varphi(z_j(s))]^2\, m(d\theta)\, ds = \\ &= \frac{1}{2} M\int_0^t\int_\Theta\iint [\varphi(z) + \tilde{f}(\theta, z, z')) + \varphi(z' + \tilde{f}(\theta, z', z)) - \\ &\qquad - \varphi(z) - \varphi(z')]^2 \mu_s(dz)\mu_s(dz')m(d\theta)ds. \end{aligned} \tag{10}$$

Making use of this calculation, we can also write the mutual character of two martingales:

$$\begin{aligned} &\langle \zeta_n(\varphi, t), \zeta_n(\psi, t)\rangle = \\ &= \frac{1}{2}\int_0^t\iint\int_\Theta [\varphi(z + \tilde{f}(\theta, z, z')) + \varphi(z' + \tilde{f}(\theta, z', z)) - \\ &- \varphi(z) - \varphi(z')][\psi(z + \tilde{f}(\theta, z, z')) + \psi(z' + \tilde{f}(\theta, z', z)) - \\ &\qquad - \psi(z) - \psi(z')]\, m(d\theta)\, \mu_s(dz)\, \mu_s(dz')\, ds. \end{aligned} \tag{11}$$

In statistical physics, to study the statistical distribution function we use the moment functions (which are sometimes called correlation functions). We set

$$m_t^{(k)}(A_1, \ldots, A_k) = M\mu_t(A_1) \ldots \mu_t(A_k).$$

It is obvious that $m_t^{(k)}(A_1, \ldots, A_k)$ is a measure with respect to each of the arguments $A_1, \ldots, A_k$. For moment functions one can obtain a system of equations in which the k-th function is expressed in terms of the (k + 1)-th (such a system is said to be 'chained'). To derive this system we shall apply Itô's formula to the product

$$\prod_{l=1}^{k}\varphi_l(z_{i_l}(t)),$$

where $\varphi_1 \in C_Z^1$, (it is the space of $\varphi \in C_Z$ with $|\varphi'| \in C_Z$), the i_1 are arbitrary indices that do not exceed n. We shall have

$$d\prod_{l=1}^{k}\varphi_l(z_{i_l}(t))=\sum_{r=1}^{l}\prod_{\substack{1\leqslant l\leqslant k\\ l\neq r}}\varphi_l(z_{i_l}(t))\Big[(\varphi_r'(z_{i_r}(t)),\ \bar{A}(z_{i_r}(t))+$$
$$+\sum_{j=1}^{n}\bar{a}(z_{i_r}(t),\ z_j(t))]\,dt+\sum_{j=1}^{n}\int_{\Theta}[\varphi_r(z_{i_r}(t)+\bar{f}(\theta,\ z_{i_r}(t),\ z_j(t))-$$
$$-\varphi_r(z_{i_r}(t))]\,p_{i_rj}(d\theta\times dt).$$

We take the mathematical expectation and sum with respect to i_l from 1 to n for all l.

We obtain

$$n^k\frac{d}{dt}\int\dots\int\varphi_1(z_1)\dots\varphi_k(z_k)\,m_t^{(k)}(dz_1,\ \dots,\ dz_k)=$$
$$=n^k\sum_{r=1}^{k}\mathsf{M}\int\dots\int\prod_{\substack{1\leqslant l\leqslant k\\ l\neq r}}\varphi_l(z_l)[(\varphi_r'(z_r),\ \bar{A}(z_r))+$$
$$+n\int\bar{a}(z_r,\ z_{k+1})\,\mu_t(dz_{k+1})+n\int\int_{\Theta}[\varphi_r(z_r+\bar{f}(\theta,\ z_r,\ z_{k+1}))-$$
$$-\varphi_r(z_r)]\,m(d\theta)\,\mu_t(dz_{k+1})\prod_{l=1}^{k}\mu_t(dz_l),$$

whence

$$\frac{d}{dt}\int\dots\int\varphi_1(z_1)\dots\varphi_k(z_k)\,m_t^{(k)}(dz_1,\ \dots,\ dz_k)=$$
$$=\sum_{r=1}^{k}\Big[\int\dots\int\prod_{\substack{1\leqslant l\leqslant k\\ l\neq r}}\varphi_l(z_l)(\varphi_r'(z_r),\ A(z_r))\,m_t^{(k)}(dz_1,\ \dots,\ dz_k)+$$
$$+n\int\dots\int\prod_{1\leqslant l\leqslant k}\varphi_l(z_l)\Big[(\varphi_r'(z_r),\ \bar{a}(z_r,\ z_{k+1}))+ \tag{12}$$
$$+\int_{\Theta}[\varphi_r(z_r+\bar{f}(\theta,\ z_r,\ z_{k+1}))-$$
$$-\varphi(z_r)]\,m(d\theta)\Big]\,m_t^{(k+1)}(dz_1,\ \dots,\ dz_{k+1})\Big].$$

If we assume that the $m_j^{(k)}(dz_1,\ \dots,\ dz_k)$ have a density relative to the Lebesgue measure:

$$m_t^{(k)}(A_1,\ \dots,\ A_k)=\int_{A_1}\dots\int_{A_k}\rho_t^{(k)}(z_1,\ \dots,\ z_k)\,dz_1\dots dz_k$$

and

$$n\int\chi_A(z+f(\theta,\ z,\ z'))\,m(d\theta)=\int_A\pi(u,\ z,\ z')\,du,$$

then we obtain from (12) the following system of equations for the densities $\rho_t^{(k)}$:

$$\frac{\partial}{\partial t}\rho_t^{(k)}(z_1,\ldots,z_k)=-\sum_{r=1}^{k}\operatorname{Sp}\frac{\partial}{\partial z_r}(\rho_t^{(k)}(z_1,\ldots,z_k)A_k(z_r))-$$
$$-n\sum_{r=1}^{k}\int\operatorname{Sp}\frac{\partial}{\partial z_r}(\rho_t^{(k+1)}(z_1,\ldots,z_{k+1})a(z_r,z_{k+1}))dz_{k+1}+$$
$$+\sum_{r=1}^{k}\iint[\rho_t^{(k+1)}(z_1,\ldots,z_{r-1},u,z_r,\ldots,z_{k+1})-$$
$$-\rho_t^{(k)}(z_1,\ldots,z_k)]\pi(u,z_r,z_{k+1})\,du\,dz_{k+1},$$

($\frac{\partial}{\partial z}(\rho A)$ and $\frac{\partial}{\partial z}(\rho a)$ are linear operators whose traces occur in the equation).

2. Problem of the Asymptotic Behavior of a Statistical Distribution Function. Compactness Conditions

The primary goal of this chapter is the study of the behavior of a system of particles when their number increases indefinitely. In this connection, the characteristics of the system (which define the coefficients of systems of stochastic equations) will depend in a definite way on n. The nature of this dependence is dictated by physical considerations. We shall consider a system of equations of the form (4), §1. The coefficient A is the force of the external field acting on a unit of mass. If the external field is invariant, then A does not vary with n. The coefficient a(x, y, v, u) is the force acting on a unit mass at the point x from the side of the particle situated at the point y. It is natural to assume that it is proportional to the mass (or, say, to the charge) of the particle. If the total mass (charge) does not vary with the growth of the number of particles, then a must be inversely proportional to n. Therefore, we shall assume that

$$a_n(x,y,v,u)=\frac{1}{n}a(x,y,v,u),$$

where a no longer depends on n (a_n is that coefficient which will occur in equation (4), §1, for n particles). The function f(θ, x, v, y, u) determines the variation of the velocity of one particle as the result of a random interaction with the other. This variation cannot depend on absolute values of masses (or charges) of particles, but only on their relations. Inasmuch as the particles are assumed to be identical, we shall also assume that the function f does not depend on n. Finally, we consider the dependence on n of the random measures $p_{ij}(d\theta\times dt)$ which define the moments and velocity jumps upon interaction of the i-th and j-th particles. We shall assume that the average number of interactions involving one particle remains approximately invariant. Inasmuch as $m(\Theta)$ is the average number of jumps of the measure $p^{ij}(d\theta \times dt)$, the average number of interactions in a unit of time of the i-th particle with the remaining particles (under the assumption that f does not vanish) will be $(n-1)m(d\theta)$. Therefore, we shall assume that

$$\mathsf{M}p_{ij}^{(n)}(d\theta \times dt) = \frac{1}{n} m(d\theta)\, dt,$$

where the measure m does not vary with n and ($p_{ij}^{(n)}$ is the Poisson measure occurring in equation (4), §1, for n particles).

If we go over to an equation of the form (6), §1, then it follows from the form of the coefficients $\tilde{A}(z)$, $\tilde{a}(z, z')$, $\tilde{f}(\theta, z, z')$ that their dependence on n is the same as that of A, a, f. Thus, we shall consider such a system of stochastic equations

$$dz_i^{(n)}(t) = \left[A(z_i^{(n)}) + \frac{1}{n} \sum_{j=1}^{n} a(z_i^{(n)}, z_j^{(n)}) \right] dt + \sum_{j=1}^{n} \int_{\Theta} f(\theta, z_i^{(n)}, z_j^{(n)})\, p_{ij}^{(n)}(d\theta \times dt) \tag{1}$$

(inasmuch as we here consider only the phase space Z, to simplify the writing we shall place a tilde, ~, over the coefficients of the equations). In the sequel we shall sometimes write z_i instead of $z_i^{(n)}$ if this does not give rise to misunderstanding. We shall denote by $\mu_t^{(n)}(dz)$ the statistical distribution function for the system (1). In the following we shall study the asymptotic behavior of the measure $\mu_t^{(n)}(dz)$ as $n \to \infty$.

We shall consider some properties of random measures in finite-dimensional Euclidean space Z. A random measure is a function $\mu(\omega, B)$ defined on $\{\Omega, \mathfrak{A}\} \times \mathfrak{B}$, where $\{\Omega, \mathfrak{A}, P\}$ is a probability space, $\mathfrak{B}$ is a σ-algebra of Borel sets in Z. The function $\mu(\omega, B)$ is measurable with respect to ω for fixed B and is a measure with respect to B for all ω. A statistical distribution function is an example of a random measure. We shall consider only normalized measures for which $\mu(\omega, Z) = 1$.

A sequence of random measures μ_n converges weakly with respect to distribtuions to the random measure μ if for every finite collection $\varphi_1(z), \ldots, \varphi_k(z)$ of functions in C_z the joint distribution of the quantities

$$\int \varphi_1(z)\, \mu_n(dz), \ \ldots, \int \varphi_k(z)\, \mu_n(dz) \tag{2}$$

converges to the joint distribtuion of the quantities

$$\int \varphi_1(z)\, \mu(dz), \ \ldots, \int \varphi_k(z)\, \mu(dz). \tag{3}$$

THEOREM 1. If for each collection $\varphi_1(z), \ldots, \varphi_k(z)$ in C_z there exists a limit distribution for the quantities (2), then there exists a random measure μ such that μ_n converges weakly with respect to distributions to μ.

<u>Proof</u>. Let $m_n(A) = M\mu_n(A)$. It is obvious that for each function $\varphi(z) \in C_z$ the limit

$$\lim_{n\to\infty} \int \varphi(z)\, m_n(dz).$$

exists. Therefore, the sequence of measures m_n converges weakly on Z to some finite measure and consequently

$$\lim_{r\to\infty} \sup_n m_n(\{z: |z| > r\}) = 0.$$

Since

$$\mathsf{P}\{\mu_n(\{z: |z| > r\}) > \varepsilon\} \leqslant \frac{1}{\varepsilon} m_n(\{z: |z| \geqslant r\}),$$

we have

$$\lim_{r\to\infty} \sup_n \mathsf{P}\{\mu_n(\{z: |z| > r\}) > \varepsilon\} = 0. \tag{4}$$

We denote by C_Z^0 the set of functions in C_Z that tend to zero at infinity. Let $\{l(\varphi, \omega),\ \varphi \in C_Z^0$ be a family of quantities on some probability space $\{\Omega, \mathfrak{A}, P\}$ such that the joint characteristic function of the magnitudes $l(\varphi_1, \omega), \ldots, l(\varphi_n, \omega)$ is defined by the relation

$$\mathsf{M}\exp\left\{i\sum_{k=1}^{m} l(\varphi_k, \omega)\right\} = \lim_{n\to\infty} \mathsf{M}\exp\left\{i\int \sum_{k=1}^{m} \varphi_k(z)\,\mu_n(dz)\right\}. \tag{5}$$

It follows from (5) that with probability 1 we have

$$l(\alpha\varphi_1 + \beta\varphi_2, \omega) = \alpha l(\varphi_1, \omega) + \beta l(\varphi_2, \omega), \tag{6}$$

$$l(\varphi, \omega) \geqslant 0 \quad \text{при} \quad \varphi \geqslant 0. \tag{7}$$

We choose a sequence of nonnegative functions $\{\psi_k \in C_Z^0\}$ such that for each $\psi \geq 0$, $\psi \in C_Z^0$,

$$\inf_k \|\psi_k - \psi\| = 0.$$

We denote by L the linear hull of the sequence $\{\psi_k\}$ with rational coefficients; that is, L is the set of functions of the form

$$\sum_{1}^{m} \alpha_i \psi_i(z),$$

where the α_i are rational. If $C \in \Omega$, then the set of those ω for which (6) is fulfilled for all $\varphi_1, \varphi_2 \in L$ and rational α and β, and also (7) for all $\varphi \in L$, $\varphi \geqq 0$, then $P(C) = 1$. For $\omega \in C$, we set

$$l^*(\varphi, \omega) = \lim_{\substack{\|\psi - \varphi\| \to 0 \\ \psi \in L}} l(\psi, \omega).$$

Then $P\{l^*(\varphi, \omega) = l(\varphi, \omega)\} = 1$ for all $\varphi \in C_Z^0$ and $l^*(\varphi, \omega) \geqq 0$ for $\varphi \geqq 0$, $\omega \in C$, $l^*(\alpha\varphi_1 + \beta\varphi_2, \omega) = \alpha l^*(\varphi_1, \omega) + \beta l^*(\varphi_2, \omega)$ for real α and β, $\varphi_1, \varphi_2 \in C_Z^0$. Consequently, for all $\omega \in C$, $l^*(\varphi, \omega)$ is a nonnegative linear functional. Thus, there exists a measure $\mu(\omega, A)$ such that, for $\varphi \in C_Z^0$,

$$l^*(\varphi, \omega) = \int \varphi(z)\mu(\omega, dz), \quad \omega \in C. \tag{8}$$

Making use of the measurability of $l^*(\varphi, \omega)$ with respect to ω for all φ, it is easy to establish that $\mu(\omega, A)$ is measurable with respect to ω for all Borel sets A. It follows from relations (4) and (5) that the sequence of random measures μ_n converges weakly to the random measure $\mu(\omega, dz)$, defined by equation (8). In fact, if $||\varphi_j|| \leqq \gamma$, $j = 1, 2, \ldots, m$, $\varphi_j \in C_Z$, and $\varphi_j^r = \varphi_j$ for $|z| \leqq r$, $\varphi_j^r \in C_Z^0$, $||\varphi_j^r|| \leqq \gamma$, then

$$\begin{aligned} &\left| \mathsf{M} \exp\left\{ i \int \sum_1^m \varphi_k(z)\mu_n(dz) \right\} - \right. \\ &\qquad \left. - \mathsf{M} \exp\left\{ i \int \sum_1^m \varphi_k(z)\mu(dz) \right\} \right| \leqslant \\ &\leqslant m\gamma \mathsf{M}\{\mu_n(\{z: |z| > r\}) + \mu(\{z: |z| > r\})\} + \\ &\quad + \left| \mathsf{M} \exp\left\{ i \int \sum_1^m \varphi_k^r(z)\mu_n(dz) \right\} - \right. \\ &\qquad \left. - \mathsf{M} \exp\left\{ i \int \sum_1^m \varphi_k^r(z)\mu(dz) \right\} \right|. \end{aligned}$$

The second term tends to zero by virtue of (5) and by a choice of r the first term can be made arbitrarily small simultaneously for all n on the basis of (4). This completes the proof of the theorem.

The sequence of random measures $\mu_n(dz)$ is said to be weakly compact with respect to a distribution if each of its subsequences contains a subsequence that converges weakly with respect to a distribution.

THEOREM 2. A necessary and sufficient condition for the sequence of random measures $\mu_n(dz)$ to be weakly compact with respect to a distribution is that the relation (4) is fulfilled for it.

Proof. Necessity. If the sequence of measures $\mu_n(dz)$ converges weakly with respect to a distribution, then the sequence of measures $m_n(dz) = M\mu_n(dz)$ will be weakly compact and hence

$$\lim_{r\to\infty}\sup_n m_n(\{z:\ |z|>r\})=0.$$

Now (4) was previously obtained from this relation in the proof of Theorem 1. This completes the proof of necessity.

Sufficiency. Suppose the set of functions L has been chosen as in Theorem 1. Inasmuch as it is countable, one can choose from any subsequence a sequence $\{n_k\}$ so that the joint distribution of the quantities $\int\varphi_1(z)\mu_{n_k}(dz)$, ..., $\int\varphi_m(z)\mu_{n_k}(dz)$ has a limit for arbitrary m and $\varphi_1, \ldots, \varphi_m \in L$. We take arbitrary $f_1, \ldots, f_m \in C_z$ and choose $\varphi_{k_1}, \ldots, \varphi_{k_m}$ in L such that

$$\sup_{|z|\leqslant r}|f_i(z)-\varphi_{k_i}(z)|\leqslant\varepsilon. \text{ Πο-}$$

Inasmuch as

$$\begin{aligned}\overline{\lim}\Big|\mathsf{M}\exp\Big\{i\sum_{j=1}^{m}\lambda_j\int f_j(z)\mu_{n_l}(dz)\Big\}-\\ -\mathsf{M}\exp\Big\{i\sum_{j=1}^{m}\lambda_j\int\varphi_{kj}(z)\mu_{n_l}(dz)\Big\}\Big|\leqslant\\ \leqslant\sum_{j=1}^{m}|\lambda_j|\big(\varepsilon+2\overline{\lim_{l\to\infty}}\,\mathsf{M}\mu_{n_l}(\{z:\ |z|\geqslant r\})\big)\end{aligned} \tag{9}$$

and the right side can be made arbitrarily small by choice of ε and r, the limit

$$\lim_{l\to\infty}\mathsf{M}\exp\Big\{i\sum_{j=1}^{n}\lambda_j\int f_j(z)\mu_{n_l}(dz)\Big\}, \tag{10}$$

which, by virtue of (9), is the uniform limit in each domain of functions, continuous in $\lambda_1, \ldots, \lambda_m$, that is, bounded with respect to $\lambda_1, \ldots, \lambda_m$. Therefore, the limit (10) will be a characteristic function. Thus, for all $f_1, \ldots, f_m \subset C_z$ the limit joint distribution of the quantities

$$\int f_1(z)\mu_{n_l}(dz), \ldots, \int f_m(z)\mu_{n_l}(dz).$$

exists. Thus, by Theorem 1, the sequence of random measures $\mu_{n_1}(dz)$ converges weakly with respect to a distribution. This completes the proof of the theorem.

COROLLARY 1. A necessary and sufficient condition for the sequence of random measures $\mu_n(dz)$ to be weakly compact with respect to a distribution is that the sequence of numerical measures $M\mu_n(dz) = m_n(dz)$ is weakly compact.

COROLLARY 2. A sufficient condition for the sequence of random measures $\mu_n(dz)$ to be weakly compact with respect to a distribution is that there exists a continuous nonnegative function $\psi(z)$ for which $\psi(z) \to \infty$ as $z \to \infty$ such that the sequence of random variables $\int\varphi(z)\mu_n(dz)$ be bounded in probability. If the sequence $\mu_n(dz)$ is weakly compact with respect to a distribution, then there exists a function $\psi(z) \geqq 0$ such that $\psi(z) \to \infty$ as $z \to \infty$ and

$$\sup_n \mathsf{M} \int \psi(z)\mu_n(dz) < \infty.$$

In fact,

$$\sup_n \mathsf{P}\{\mu_n(\{z: |z| > r\}) > \varepsilon\} \leqslant$$
$$\leqslant \sup_n \mathsf{P}\left\{\frac{1}{\inf\limits_{|z|>r} \psi(z)} \int \psi(z)\mu_n(dz) > \varepsilon\right\} =$$
$$= \sup_n \mathsf{P}\left\{\int \psi(z)\mu_n(dz) > \varepsilon \inf_{|z|>r} \psi(z)\right\} \to 0$$

as $r \to \infty$ for all $\varepsilon > 0$. Furthermore, it follows from the compactness of the measures $m_n(dz)$ that there exists a sequence $r_k \to \infty$ such that $m_n(\{z: |z| > r_k\}) \leqq 2^{-k}$. If $\psi(z)$ is a continuous function such that $k \leqq \psi(z) \leqq k + 1$ for $r_k \leqq |z| \leqq r_{k+1}$, then

$$\int \psi(z)m_n(dz) \leqslant \sum_k (k+1)2^{-k}.$$

COROLLARY 3. Suppose given k sequences of random measures $\{\mu_n^1, \ldots, \mu_n^k\}$ each of which is weakly compact with respect to a distribution. Then the sequence of random measures $\mu_n^1 \times \mu_n^2 \times \ldots \times \mu_n^k \in Z^k$ is also weakly compact with respect to a distribution.

This follows from the weak compactness of the measure

$$\begin{aligned} M\mu_n^1(dz_1)\,\mu_n^2(dz_2)\ldots\mu_n^k(dz_k) &\leqslant \\ \leqslant (M|\mu_n^1(dz_1)|^k)^{1/k} \times \ldots \times (M|\mu_n^k(dz_k)|^k)^{1/k} &\leqslant \\ \leqslant (M\mu_n^1(dz_1)\ldots M\mu_n^k(dz_k))^{1/k} & \end{aligned}$$

(we made use of the fact that $0 \leqq \mu_n^i \leqq 1$).

3. Limit Theorem for Steplike Processes

In this section we consider a system of particles with the absence of an external field and of remote interaction between particles. Between the particles there are only random forces which momentarily change the position of the particles. It is also assumed that the particles are fixed between moments of interaction. Of course, from the physical viewpoint, systems of particles with such interactions are meaningless. However, as we shall see in the sequel, such artificial systems approximate sufficiently well the systems that were described above. On the other hand, systems with jumplike variations are simpler to investigate from the mathematical point of view.

Thus, suppose given a system of stochastic equations

$$dz_i(t) = \sum_{j=1}^{n} \int_{\Theta} f(\theta, z_i, z_j)\, p_{ij}^{(n)}(d\theta \times dt), \quad i = 1, \ldots, n \tag{1}$$

with the initial conditions $z_i(0) = z_i^0$.

THEOREM 1. Suppose the following three conditions are fulfilled:

(1) $Mp_{ij}^{(n)}(d\theta \times dt) = \frac{1}{n}m(d\theta)dt$, where m is a finite measure;

(2) there exists an L such that

$$\int |f(\theta, z_1, z_2) - f(\theta, z_1', z_2')|\, m(d\theta) \leqslant L(|z_1 - z_1'| + |z_2 - z_2'|),$$

$$\int |f(\theta, z_1, z_2)|\, m(d\theta) \leqslant L(1 + |z_1| + |z_2|);$$

(3) for each $\varphi \in C_Z$ the limit

$$\lim_{n\to\infty} \int \varphi(z)\, \mu_0^{(n)}(dz) = \lim_{n\to\infty} \frac{1}{n} \sum_{i=1}^{n} \varphi(z_i^0) = \int \varphi(z)\, \lambda_0(dz),$$

exists, where $\lambda_0(dz)$ is a nonrandom measure on Z.

Then, for all $t > 0$, the sequence of measures $\mu_t^{(n)}(dz)$ converges weakly with respect to a distribution to the nonrandom measure $\lambda_t(dz)$, which is the unique solution of the equation: for all $\varphi \in C_Z$

$$\frac{d}{dt}\int \varphi(z)\lambda_t(dz) = \\ = \iiint\limits_{\Theta} [\varphi(z + f(\theta, z, z')) - \varphi(z)]\, m(d\theta)\,\lambda_t(dz)\,\lambda_t(dz'),$$

for which for $\varphi \in C_Z$ the function $\int\varphi(z)\lambda_t(dz)$ is continuous for $t \geq 0$ and coincides with $\int\varphi(z)\lambda_0(dz)$ for $t = 0$.

Proof. 1. We first establish that (2) has a unique solution with the required properties. Let $\bar\lambda_t(dz)$ be some other solution for which $\bar\lambda_0(dz) = \lambda_0(dz)$ and $\bar\lambda_t(dz)$ is continuous. Writing (2) for $\bar\lambda_t(dz)$, subtracting one equation from the other and integrating with respect to t from 0 to t, we obtain

$$\int \varphi(z)\lambda_t(dz) - \int \varphi(z)\bar\lambda_t(dz) = \\ = \int_0^t \iiint\limits_{\Theta} [\varphi(z + f(\theta, z, z')) - \varphi(z)]\, m(d\theta) \times \\ \times \lambda_s(dz)[\lambda_s(dz') - \bar\lambda_s(dz')]\, ds + \\ + \int_0^t \iiint\limits_{\Theta} [\varphi(z + f(\theta, z, z')) - \varphi(z)] \times \\ \times m(d\theta)\,\bar\lambda_s(dz')[\lambda_s(dz) - \bar\lambda_s(dz)]\, ds.$$

We denote by L_1 the set of functions $\varphi \in C_Z$ for which $||\varphi|| \leq 1$ and $|\varphi(z) - \varphi(z')| \leq |z - z'|$. Let $\varphi \in L_1$. We consider the functions

$$\psi_1(z') = \iint\limits_{\Theta} [\varphi(z + f(\theta, z, z')) - \varphi(z)]\, m(d\theta)\,\lambda_t(dz),$$

$$\psi_2(z) = \iint\limits_{\Theta} [\varphi(z + f(\theta, z, z')) - \varphi(z)]\, m(d\theta)\,\lambda_t(dz').$$

Obviously,

$$|\psi_k(z)| \leqslant 2m(\Theta),$$

$$|\psi_1(z') - \psi_1(\bar z')| \leqslant \int |f(\theta, z, z') - f(\theta, z, \bar z')|\, m(d\theta)\,\lambda_t(dz),$$

$$|\psi_2(z) - \psi_2(\bar z)| \leqslant \int |f(\theta, z, z') - f(\theta, \bar z, z')|\, m(d\theta)\,\lambda_t(dz') + \\ + |z - \bar z|\, m(\Theta) + |\varphi(z) - \varphi(\bar z)|\, m(\Theta),$$

hence,

$$|\psi_k(z) - \psi_k(\bar{z})| \leqslant (L + 2m(\Theta))|z - \bar{z}|.$$

Therefore, for $\lambda = \frac{1}{L+2m(\Theta)}$, the functions $\lambda\psi_1(z)$ and $\lambda\psi_2(z)$ belong to L_1. Consequently,

$$\sup_{\varphi \in L_1} \left| \int \varphi(z)\lambda_t(dz) - \int \varphi(z)\bar{\lambda}_t(dz) \right| \leqslant$$
$$\leqslant \frac{1}{2\lambda} \int_0^t \sup_{\psi \in L_1} \left| \int \psi(z)\lambda_s(dz) - \int \psi(z)\bar{\lambda}_s(dz) \right| ds.$$

From this it follows that $\int\varphi(z)\lambda_t(dz) = \int\varphi(z)\bar{\lambda}_t(dz)$ for $\varphi \in L_1$. But then the measures λ_t and $\bar{\lambda}_t$ coincide. The uniqueness of the solution of equation (2) is thus proven.

2. We shall show that for all t the sequence of random measures $\mu_t^{(n)}$ is weakly compact with respect to a distribution. Let $\psi(\lambda)$, $\lambda > 0$, be a function having the following properties:

(1) $\psi(\lambda)$ is continuous, nonnegative and $\psi(\lambda) \to \infty$ as $\lambda \to \infty$,
(2) $\psi(\lambda)$ is subadditive: $\psi(\alpha+\beta) \leqq \psi(\alpha) + \psi(\beta)$ for $\alpha > 0$, $\beta > 0$,
(3) $\psi(\lambda)$ is concave upward,
(4) $\overline{\lim\limits_{n\to\infty}} \int\psi(|z|)\mu_0^{(n)}(dz) < \infty$.

Then

$$d\int \psi(|z|)\mu_t^{(n)}(dz) =$$
$$= \sum_{i,j=1}^{n} \frac{1}{n} \int [\psi(|z_i + f(\theta, z_i, z_j)|) - \psi(|z_i|)]\, p_{ij}^{(n)}(d\theta \times dt).$$

Using the properties of ψ, we can write:

$$\int [\psi(|z_i + f(\theta, z_i, z_j)|) - \psi(|z_i|)]\, m(d\theta) \leqslant$$
$$\leqslant \int \psi(|f(\theta, z_i, z_j)|)\, m(d\theta) = m(\Theta) \int \psi(|f(\theta, z_i, z_j)|) \frac{m(d\theta)}{m(\Theta)} \leqslant$$
$$\leqslant m(\Theta)\, \psi\left(\int |f(\theta, z_i, z_j)| \frac{m(d\theta)}{m(\Theta)}\right) \leqslant$$
$$\leqslant m(\Theta)\, \psi\left(\frac{L}{m(\Theta)}(1 + |z_i| + |z_j|)\right).$$

Obviously, the function $\psi(\lambda)$ is increasing. If r is an integer such that $\frac{L}{m(\Theta)} \leqq r$, then

$$\psi\left(\frac{L}{m(\Theta)}(1 + |z_i| + |z_j|)\right) \leqslant r(\psi(1) + \psi(|z_i|) + \psi(|z_j|)).$$

Therefore,

$$\mathsf{M}d\int \psi(|z|)\,\mu_t^{(n)}(dz) = \\ = \sum_{i,j=1}^{n} \frac{1}{n^2}\int_{\Theta} [\psi(z_i + f(\theta, z_i, z_j)) - \psi(z_i)]\, m(d\theta)\, dt \leqslant \\ \leqslant 2rm(\Theta)\int \psi(|z|)\,\mu_t^{(n)}(dz)\,dt + r\psi(1)\,m(\Theta)\,dt.$$

From this it follows that

$$\mathsf{M}\int \psi(|z|)\,\mu_t^{(n)}(dz) \leqslant \left[\int \psi(|z|)\,\mu_t^{(0)}(dz) + c\right]\exp\{2rm(\Theta)t\},$$

where c is some constant. Thus,

$$\varlimsup_{n\to\infty} \mathsf{M}\int \psi(|z|)\,\mu_t^{(n)}(dz) < \infty.$$

By Corollary 2, §2, we conclude that $\mu_t^{(n)}$ is a sequence of measures that is compact with respect to distributions.

3. It follows from Corollary 3, §2, that for any finite collection $t_1, \dots, t_m$ the measures $\mu_{t_1}^{(n)} \times \dots \times \mu_{t_m}^{(n)}$ will also be compact. Therefore, for all $t_1, \dots, t_m$, we can select a subsequence n_i so that for arbitrary functions $\varphi_1(z), \dots, \varphi_m(z) \in C_Z$ the joint distribution

$$\int \varphi_1(z)\,\mu_{t_1}^{(n)}(dz), \dots, \int \varphi_m(z)\,\mu_{t_m}^{(n)}(dz) \tag{4}$$

converges to the joint distribution of the variables

$$\int \varphi_1(z)\,\nu_{t_1}(dz), \dots, \int \varphi_m(z)\,\nu_{t_m}(dz), \tag{5}$$

where $\nu_{t_1}, \dots, \nu_{t_m}$ are certain random measures. Using the Cantor diagonal method, we convince ourselves that there exist a subsequence n_l and random measures $\nu_t(dz)$ defined for all rational t such that the joint distribution of the variables (4) converges to the joint distribution of the variables (5) for arbitrary rational numbers $t_1, \dots, t_m$ and $\varphi_k \in C_Z$. We now note that (1) implies the following relation: $h > 0$,

$$\int \varphi(z)\,\mu_{t+h}^{(n)}(dz) - \int \varphi(z)\,\mu_t^{(n)}(dz) =$$
$$= \frac{1}{n} \sum_{i,j=1}^{n} \int_t^{t+h} \int_\Theta [\varphi(z_i + f(\theta, z_i, z_j)) - \varphi(z_i)]\, p_{ij}^{(n)}(d\theta \times ds).$$

Consequently,

$$\mathsf{M}\left|\int \varphi(z)\,\mu_{t+h}^{(n)}(dz) - \int \varphi(z)\,\mu_t^{(n)}(dz)\right| \leqslant$$
$$\leqslant \frac{1}{n^2} \int_t^{t+h} \int_\Theta \sum_{i,j=1}^{n} |\varphi(z_i + f(\theta, z_i, z_j)) - \varphi(z_i)|\, m(d\theta)\, ds \leqslant \tag{6}$$
$$\leqslant 2\|\varphi\|\, m(\Theta)\, h.$$

Therefore, substituting n_1 for n in this relation and then passing to the limit, we convince ourselves that for all rational t and h,

$$\mathsf{M}\left|\int \varphi(z)\,\nu_{t+h}(dz) - \int \varphi(z)\,\nu_t(dz)\right| \leqslant 2\|\varphi\|\, m(\Theta)\, h.$$

It follows from the preceding inequality that for all irrational t and for all $\varphi \in C_Z$ the limit exists in the sense of mean convergence:

$$\lim_{s\to t} \int \varphi(z)\,\nu_s(dz),$$

where s assumes rational values. We denote this limit by $\nu_t[\varphi]$.

It is obvious that the inequality

$$\mathsf{M}\,|\nu_{t+h}[\varphi] - \nu_t[\varphi]| \leqslant 2\|\varphi\|\, m(\Theta)\, h.$$

is valid for $\nu_t[\varphi]$. Using inequalityies (6) and (8), it is easy to convince oneself that the joint distribution of the variables (4) converges for all $t_1, t_2, \ldots, t_m \geq 0$ and $\varphi_1, \ldots, \varphi_m \in C_Z$ to the joint distribution of the variables $\nu_{t_1}[\varphi_1], \ldots, \nu_{t_m}[\varphi_m]$. Therefore, by Theorem 1, §2, there exist measures ν_t such that for all rest $t \geq 0$ we have $\nu_t[\varphi] = \int\varphi(z)\nu_t(dz)$.

4. We now write relation (8), §1, for $\mu_t^{(n)}dz$ taking into consideration that $\tilde{A} = 0$, $\tilde{a} = 0$, and instead of the measure $m(d\theta)$ one must write $\frac{1}{n}m(d\theta)$. We then have

$$d\int \varphi(z)\,\mu_t^{(n)}(dz) =$$
$$= \iint\int_\Theta [\varphi(z + f(\theta, z, z')) - \varphi(z)]\, m(d\theta)\,\mu_t^{(n)}(dz)\,\mu_t^{(n)}(dz') +$$
$$+ \frac{1}{n}\sum_{i,j=1}^{n}\int_\Theta [\varphi(z_i + f(\theta, z_i, z_j)) - \varphi(z_i)]\, q_{ij}^{(n)}(d\theta \times dt).$$

Integrating this equation with respect to t from 0 to t we obtain

$$\int \varphi(z)\,\mu_t^{(n)}(dz) = \int \varphi(z)\,\mu_0^{(n)}(dz) +$$
$$+ \int_0^t \iint\int_\Theta [\varphi(z + f(\theta, z, z')) - \varphi(z)]\, m(d\theta)\,\mu_s^{(n)}(dz)\,\mu_s^{(n)}(dz')\,ds +$$
$$+ \frac{1}{n}\sum_{i,j=1}^{n}\int_0^t\int_\Theta [\varphi(z_i + f(\theta, z_i, z_j)) - \varphi(z_i)]\, q_{ij}^{(n)}(d\theta \times dt).$$

We denote

$$\int [\varphi(z + f(\theta, z, z')) - \varphi(z)]\, m(d\theta) = \Phi(z, z');$$

this is a continuous, bounded function of two variables,

$$\sup_{z, z'} |\Phi(z, z')| \leqslant 2m(\Theta)\|\varphi\|.$$

Inasmuch as

$$\iint \Phi(z, z')\,\mu_s^{(n)}(dz)\,\mu_s^{(n)}(dz') = \frac{1}{n^2}\sum_{i,j=1}^{n} \Phi(z_i(s), z_j(s)),$$

we have by Itô's formula that

$$d_s \iint \Phi(z, z')\,\mu_s^{(n)}(dz)\,\mu_s^{(n)}(dz') =$$
$$= \frac{1}{n^3}\sum_{i,j,k}\Big[\int_\Theta \{\Phi(z_i(s) + f(\theta, z_i(s), z_k(s), z_j(s)) -$$
$$- \Phi(z_i(s), z_j(s))\}\, p_{ik}^{(n)}(d\theta \times ds) +$$
$$+ \int_\Theta \{\Phi(z_i(s), z_j(s) + f(\theta, z_j(s), z_k(s))) -$$
$$- \Phi(z_i(s), z_j(s))\}\, p_{jk}^{(n)}(d\theta \times ds)\Big] +$$
$$+ \frac{1}{n^2}\sum_{i,j}\int_\Theta [\Phi(z_i(s) + f(\theta, z_i(s), z_j(s)), z_j(s) -$$
$$- f(\theta, z_i(s), z_j(s))) - \Phi(z_i(s), z_j(s))]\, p_{ij}^{(n)}(d\theta \times ds).$$

Therefore,

$$\mathsf{M}\left|\iint \Phi(z, z')\mu_{t+h}^{(n)}(dz)\mu_{t+h}^{(n)}(dz') - \right.$$
$$\left. - \iint \Phi(z, z')\mu_t^{(n)}(dz)\mu_t^{(n)}(dz')\right| \leqslant \tag{9}$$
$$\leqslant 2\sup_{z, z'}|\Phi(z, z')|\left(2+\frac{1}{n}\right)m(\Theta)h \leqslant 12m^2(\Theta)\|\varphi\| h.$$

Making use of this estimate, we convince ourselves that

$$\mathsf{M}\left|\int_0^t \iint \Phi(z, z')\mu_s^{(n)}(dz)\mu_s^{(n)}(dz')\,ds - \right.$$
$$\left. - \sum_{k=1}^{r}\frac{t}{r}\iint \Phi(z, z')\mu_{kt/r}^{(n)}(dz)\mu_{kt/r}^{(n)}(dz')\right| \leqslant$$
$$\leqslant \sum_{k=1}^{r}\int_{(k-1)t/r}^{kt/r}\left|\iint \Phi(z, z')\mu_s^{(n)}(dz)\mu_s^{(n)}(dz') - \right. \tag{10}$$
$$\left. - \iint \Phi(z, z')\mu_{kt/r}^{(n)}(dz)\mu_{kt/r}^{(n)}(dz')\right| ds \leqslant$$
$$\leqslant \sum_{k=1}^{r}\int_{(k-1)t/r}^{kt/r}\left(\frac{kt}{r}-s\right)ds\cdot 12m^2(\Theta)\|\varphi\| = 6m^2(\Theta)\|\varphi\|\frac{t^2}{r}.$$

5. We shall show that for all continuous bounded functions $\phi(z, z')$ on Z^2 and $t_1, \ldots, t_m \geqq 0$ the joint distribution of the variables

$$\iint \Phi(z, z')\mu_{t_1}^{(nt)}(dz)\mu_{t_1}^{(nt)}(dz'), \ldots$$
$$\ldots, \iint \Phi(z, z')\mu_{t_m}^{(nt)}(dz)\mu_{t_m}^{(nt)}(dz') \tag{11}$$

converges to the joint distribution of the variables

$$\iint \Phi(z, z')\nu_{t_1}(dz)\nu_{t_1}(dz'), \ldots, \iint \Phi(z, z')\nu_{t_m}(dz)\nu_{t_m}(dz'). \tag{12}$$

We now assume that

$$\Phi(z, z') = \sum_{k=1}^{N}\varphi_k(z)\psi_k(z'), \quad \varphi, \psi \in C_Z. \tag{13}$$

Then, using the relation

$$\iint \Phi(z, z')\mu_{t_i}^{(nt)}(dz)\mu_{t_i}^{(nt)}(dz') =$$
$$= \sum_{k=1}^{N}\int \varphi_k(z)\mu_{t_i}^{(nt)}(dz)\int \psi_k(z')\mu_{t_i}^{(nt)}(dz')$$

and the convergence of the joint distribution of the variables

$$\left\{\int \varphi_k(z)\mu_{t_i}^{(n_l)}(dz),\ \int \psi_k(z)\mu_{t_i}^{(n_l)}(dz),\ k=1,\ldots,N;\ i=1,\ldots,m\right\}$$

to the joint distribution of the variables

$$\left\{\int \varphi_k(z)\nu_{t_i}(dz),\ \int \psi_k(z)\nu_{t_i}(dz),\ k=1,\ldots,N;\ i=1,\ldots,m\right\},$$

we convince ourselves that teh joint distribution of the variables (11) converges to the joint distribution of the variables (12). But for arbitrary $\Phi(z, z')$ and for each $\varepsilon > 0$ one can select a function $\Phi_\varepsilon(z, z')$ of the form (13) such that

$$\mathsf{M}\left|\iint \Phi(z, z')\mu_{t_i}^{(n_l)}(dz)\mu_{t_i}^{(n_l)}(dz') - \right.$$
$$\left. - \iint \Phi_\varepsilon(z, z')\mu_{t_i}^{(n_l)}(dz)\mu_{t_i}^{(n_l)}(dz')\right| \leqslant \varepsilon,$$
$$\mathsf{M}\left|\iint \Phi(z, z')\nu_{t_i}(dz)\nu_{t_i}(dz') - \right.$$
$$\left. - \iint \Phi_\varepsilon(z, z')\nu_{t_i}(dz)\nu_{t_i}(dz')\right| \leqslant \varepsilon$$

for all $i = 1, \ldots, m$. (For this it is necessary that $|\Phi(z, z') - \Phi(z, z')| < \varepsilon/2$ for $|z| \leqq \alpha$, $|z'| \leqq \alpha$, and

$$\sup_{n_l, i}\sup_{z, z'}[|\Phi(z, z')| + |\Phi_\varepsilon(z, z')|] \times$$
$$\times \left(\mathsf{M}\mu_{t_i}^{(n_l)}(\{z : |z| > \alpha\}) + \mathsf{M}\nu_{t_i}(\{z : |z| > \alpha\})\right) \leqslant \varepsilon/4.)$$

The assertion made in this subsection follows from the convergence of the joint distribution of the variables (11) when Φ is replaced by Φ_ε to the joint distribution of the variables (12) (when again Φ is replaced by Φ_ε) and the inequalities (14).

6. Using estimate (10) and the result of the preceding subsection we can convince ourselves that the joint distribution of the variables

$$\sum_{k=1}^{r} \frac{t}{r} \iint \Phi(z, z')\mu_{kt/r}^{(n_l)}(dz)\mu_{kt/r}^{(n_l)}(dz')$$

and

$$\sum_{k=1}^{q} \frac{t}{q} \iint \Phi(z, z')\mu_{kt/q}^{(n_l)}(dz)\mu_{kt/q}^{(n_l)}(dz')$$

converges as $l \to \infty$ to the joint distribution of the variables

$$\sum_{1}^{r}\frac{t}{r}\iint\Phi(z,\ z')\,\nu_{kt/r}(dz)\,\nu_{kt/r}(dz')$$

and

$$\sum_{k=1}^{q}\frac{t}{q}\iint\Phi(z,\ z')\,\nu_{kt/q}(dz)\,\nu_{kt/q}(dz') \tag{15}$$

and

$$\mathsf{M}\left|\sum_{k=1}^{r}\frac{t}{r}\iint\Phi(z,\ z')\,\nu_{kt/r}(dz)\,\nu_{kt/r}(dz')-\right.$$
$$\left.-\sum_{k=1}^{q}\frac{t}{r}\iint\Phi(z,\ z')\,\nu_{kt/q}(dz)\,\nu_{kt/q}(dz')\right|\leqslant$$
$$\leqslant 6m^2(\Theta)\|\varphi\|t^2\left(\frac{1}{r}+\frac{1}{q}\right).$$

On the other hand, it follows from estimate (9) and the result in subsection 5 that

$$\iint\Phi(z,\ z')\,\nu_s(dz)\,\nu_s(dz')$$

is stochastically continuous with respect to s; therefore,

$$\lim_{q\to\infty}\sum_{k=1}^{q}\frac{t}{q}\iint\Phi(z,\ z')\,\nu_{kt/q}(dz)\,\nu_{kt/q}(dz')=$$
$$=\int_0^t\iint\Phi(z,\ z')\,\nu_s(dz)\,\nu_s(dz')\,ds$$

in the sense of convergence in probability. Therefore, passing in (15) to the limit as $q\to\infty$, we obtain

$$\mathsf{M}\left|\sum_{k=1}^{r}\frac{t}{r}\iint\Phi(z,\ z')\,\nu_{kt/r}(dz)\,\nu_{kt/r}(dz')-\right.$$
$$\left.-\int_0^t\iint\Phi(z,\ z')\,\nu_s(dz)\,\nu_s(dz')\,ds\right|\leqslant 6m^2(\Theta)\|\varphi\|\frac{t^2}{r}. \tag{16}$$

From the estimates (10) and (16), the fact that r is arbitrary and the convergence of the joint distribution of the variables

$$\int \varphi(z)\,\mu_t^{(n_l)}(dz) \quad \text{и} \quad \sum_{k=1}^{r} \frac{t}{r} \iint \Phi(z, z')\,\mu_{kt/r}^{(n_l)}(dz)\,\mu_{kt/r}^{(n_l)}(dz')$$

to the joint distribution of the variables

$$\int \varphi(z)\,\nu_t(dz) \quad \text{и} \quad \int_0^t \iint \Phi(z, z')\,\nu_s(dz)\,\nu_s(dz')\,ds$$

and the existence of the nonrandom limit

$$\lim_{l\to\infty} \int \varphi(z)\,\mu_0^{(n_l)}(dz) = \int \varphi(z)\,\lambda_0(dz),$$

and also the estimate

$$\mathsf{M}\left(\frac{1}{n} \sum_{i,j=1}^{n} \int_\Theta [\varphi(z_i + f(\theta, z_i, z_j)) - \varphi(z_i)]\,q_{ij}^{(n)}(d\theta \times dt)\right)^2 = \\ = O\left(\frac{1}{n}\right)$$

we obtain from relation (8) the following equation

$$\int \varphi(z)\,\nu_t(dz) = \int \varphi(z)\,\lambda_0(dz) + \int_0^t \iiint_\Theta [\varphi(z + f(\theta, z, z')) - \\ - \varphi(z)]\,m(d\theta)\,\nu_s(dz)\,\nu_s(dz')\,ds. \tag{17}$$

The relation (17) is fulfilled for each $\varphi \in C_Z$ and $t > 0$ with probability 1. It is easy to see that if it is fulfilled for some countable set of functions in $\overset{\circ}{C}_Z$, dense in $\overset{\circ}{C}_Z$, and all rational t, then it is also fulfilled for all φ in C_Z and $t > 0$. Therefore, there exists a set $\Lambda \in \Omega$ such that $P(\Lambda) = 1$ and the relation (17) is valid for all $\varphi \in C_Z$ for $\omega \in \Lambda$. For $\omega \in \Lambda$, $\int\varphi(z)\nu_t(dz)$ is a continuous function of t since the left member of (17) is continuous. If $\omega \in \Lambda$ and $\omega_1 \in \Lambda$, then it follows from the uniqueness of the solution of equation (2) that $\nu_t(\omega, dz) = \nu_t(\omega_1, dz)$. Consequently, for $\omega \in \Lambda$, $\nu_t(\omega, dz)$ does not depend on ω and coincides with the unique solution of equation (2) (that the solution of (2) exists follows from the fact that Λ is not empty since $P(\Lambda) = 1$).

7. We denote by $\lambda_t(dz)$ the unique solution of equation (2). We shall show that $\int\varphi(z)\mu_t^{(n)}(dz)$ as $n \to \infty$ converges in probability to $\int\varphi(z)\lambda_t(dz)$ (inasmuch as the preceding variable is nonrandom, con-

vergence in probability is equivalent to convergence in distribution). If this is not so, then for some function $\varphi \in C_Z$ oen can find a sequence n_l such that for some $\varepsilon > 0$

$$\mathsf{P}\left\{\left|\int \varphi(z)\,\mu_t^{(n_l)}(dz) - \int \varphi(z)\,\lambda_t(dz)\right| > \varepsilon\right\} > \varepsilon \tag{18}$$

for all l. As follows from subsections 3–6, one can assume that this sequence n_l is such that the measures $\mu_t^{(n_l)}(dz)$ converges weakly with respect to distributions to certain measures $\nu_t(dz)$ for which (17) is fulfilled. But then $\nu_t(dz) = \lambda_t(dz)$ and

$$\lim_{n\to\infty} \int \varphi(z)\,\mu_t^{(n_l)}(dz) = \int \varphi(z)\,\lambda_t(dz),$$

which contradicts (18). This completes the proof of the theorem.

The case when the number of interactions on one particle increases to infinity as n increases is of interest. In order to describe such a situation one can make use of equation (1) for which the measure $m(d\theta)$ is infinite but the function $f(\theta, z, z')$ depends on n in such a way that it is different from zero only on the set $\Theta_n \subset \Theta$ and $m(\Theta_n) < \infty$. Therefore, we shall consider equations of the form (1) in which instead of the function $f(\theta, z, z')$ there figures $f_n(\theta, z, z') = \chi_{\Theta_n}(\theta) f(\theta, z, z')$, where f no longer depends on n. Then it can be rewritten in the following form:

$$dz_i^{(n)}(t) = \sum_{j=1}^{n} \int_{\Theta_n} f(\theta, z_i^{(n)}, z_j^{(n)})\, p_{ij}^{(n)}(d\theta \times dt). \tag{1'}$$

We shall assume that the 'total' action in a unit of time remains bounded; that is,

$$\sup_n \int_{\Theta_n} |f(\theta, z, z')| \times m(d\theta) < \infty.$$

It is natural to assume that $\Theta_n \subset \Theta_{n+1}$ and $\Theta = \cup\Theta_n$. Then the preceding condition is equivalent to the condition

$$\int_{\Theta} |f(\theta, z, z')|\, m(d\theta) < \infty.$$

We investigate the conditions for weak compactness with respect to a distribution for statistical distribution functions of equation (1'). Let $\psi(z)$ be a positive subadditive function for $z \neq 0$ that is equal to

$|z|$ in some neighborhood of the point 0 and which tends to ∞ as $z \to \infty$;

$$\mathsf{M}\int \psi(z)\,\mu_t^{(n)}(dz) = \mathsf{M}\int \psi(z)\,\mu_0^{(n)}(dz) + \\ + \mathsf{M}\int_0^t \int\int\int_{\Theta_n} [\psi(z + f(\theta,\ z,\ z')) - \psi(z)] \times \\ \times m(d\theta)\,\mu_s^{(n)}(dz)\,\mu_s^{(n)}(dz')\,ds \leqslant \mathsf{M}\int \psi(z)\,\mu_0^{(n)}(dz) + \\ + \mathsf{M}\int_0^t \int\int\int_{\Theta_n} \psi(f(\theta,\ z,\ z'))\,m(d\theta)\,\mu_s^{(n)}(dz)\,\mu_s^{(n)}(dz')\,ds.$$

Suppose there exists a function $\psi(z)$ with the indicated properties such that for some c we have

$$\int_{\Theta} \psi(f(\theta,\ z,\ z'))\,m(d\theta) \leqslant c(1 + \psi(z) + \psi(z')). \tag{19}$$

Then

$$\mathsf{M}\int \psi(z)\,\mu_t^{(n)}(dz) \leqslant \left[\mathsf{M}\int \psi(z)\,\mu_0^{(n)}(dz) + ct\right] + \\ + 2c\int_0^t \mathsf{M}\int \psi(z)\,\mu_s^{(n)}(dz)\,ds$$

and

$$\mathsf{M}\int \psi(z)\,\mu_t^{(n)}(dz) \leqslant \left[\mathsf{M}\int \psi(z)\,\mu_0^{(n)}(dz) + ct\right] e^{2ct}. \tag{20}$$

Thus, for the weak compactness with resepct to a distribution of the sequence $\mu_t^{(n)}$ it suffices to assume that

$$\mathsf{M}\int \psi(z)\quad \mu_0^{(n)}(dz)$$

is bounded and that condition (19) is fulfilled.

To investigate the asymptotic behavior of $\mu_t^{(n)}$ we consider, together with the system of equation (1'), the system

$$dz_i^{(n,m)}(t) = \sum_{i,j=1}^{n} \int_{\Theta_m} f(\theta,\ z_i^{(n,m)},\ z_j^{(n,m)})\,p_{ij}^{(n)}(d\theta \times dt) \tag{21}$$

with the same initial conditions as for the system (1). Here, m will be fixed, n > m. We denote the statistical distribution function for this system by $\mu_t^{(n,m)}(dz)$. We estimate the difference between the integrals with respect to $\mu_t^{(n)}$. We have

$$d\psi(z_i^{(n)}(t) - z_i^{(n,m)}(t)) =$$
$$= \sum_{j=1}^{n} \int_{\Theta_m} [\psi(z_i^{(n)} + f(\theta, z_i^{(n)}, z_j^{(n)}) - z_i^{(n,m)} -$$
$$- f(\theta, z_i^{(n,m)}, z_j^{(n,m)})) - \psi(z_i^{(n)} - z_i^{(n,m)})\, p_{ij}^{(n)}(d\theta \times dt) +$$
$$+ \sum_{j=1}^{n} \int_{\Theta_n \setminus \Theta_m} [\psi(z_i^{(n)} + f(\theta, z_i^{(n)}, z_j^{(n)}) - z_i^{(n,m)}) -$$
$$- \psi(z_i^{(n)} - z_i^{(n,m)})]\, p_{ij}^{(n)}(d\theta \times dt).$$

Integrating this relation with respect to time, taking into account that $z_i^{(n)}(0) = z_i^{(n,m)}(0)$ and computing the mathematical expectation, and then making use of teh subadditivity of ψ, we obtain

$$\mathsf{M}\psi(z_i^{(n)}(t) - z_i^{(n,m)}(t)) \leqslant$$
$$\leqslant \frac{1}{n} \sum_{j=1}^{n} \int_0^t \int_\Theta \mathsf{M}\psi(f(\theta, z_i^{(n)}(s), z_j^{(n)}(s)) -$$
$$- f(\theta, z_i^{(n,m)}(s), z_j^{(n,m)}(s)))\, m(d\theta)\, ds +$$
$$+ \frac{1}{n} \sum_{j=1}^{n} \int_0^t \int_{\Theta_n \setminus \Theta_m} \mathsf{M}\psi(f(\theta, z_i^{(n)}(s), z_j^{(n)}(s)))\, m(d\theta)\, ds.$$

Suppose

$$\sup_{z, z'} \frac{1}{1 + \psi(z) + \psi(z')} \int_{\Theta \setminus \Theta_m} \psi(f(\theta, z, z'))\, m(d\theta) = \varepsilon_m \to 0 \tag{22}$$

and that there exists a constant L for which

$$\int_\Theta \psi(f(\theta, z, z') - f(\theta, \bar{z}, \bar{z}'))\, m(d\theta) \leqslant$$
$$\leqslant L[\psi(z - \bar{z}) + \psi(z' - \bar{z}')]. \tag{23}$$

Then

$$\frac{1}{n} \sum_{i=1}^{n} \mathsf{M}\psi(z_i^{(n)}(t) - z_i^{(n,m)}(t)) \leqslant$$
$$\leqslant 2L \frac{1}{n} \sum_{i=1}^{n} \int_0^t \mathsf{M}\psi(z_i^{(n)}(s) - z_i^{(n,m)}(s))\, ds +$$

$$+\varepsilon_m \int_0^t \left(1 + 2\int \psi(z)\,\mu_s^{(n)}(dz)\right)$$

Consequently,

$$\frac{1}{n}\sum_{i=1}^{n} \mathsf{M}\psi(z_i^{(n)}(t) - z_i^{(n,m)}(t)) \leqslant$$

$$\leqslant \varepsilon_m e^{2Lt}\left[t + 2\int_0^t \int \psi(z)\,\mu_s^{(n)}(dz)\,ds\right].$$

Making use of estimate (20) we convince ourselves that

$$\frac{1}{n}\sum_{i=1}^{n} \mathsf{M}\psi(z_i^{(n)}(t) - z_i^{(n,m)}(t)) = O(\varepsilon_m)$$

uniformly with respect to n and on each finite interval of time provided

$$\overline{\lim_{n\to\infty}}\, \mathsf{M}\int \psi(z)\,\mu_0^{(n)}(dz) < \infty.$$

Let $\varphi(z) \in C_Z$ and $|\varphi(z_1) - \varphi(z_2)| \leq K|z_1 - z_2|$. Then, for some K_1, we have $|\varphi(z_1) - \varphi(z_2)| \leq K_1\psi(z_1 - z_2)$. Therefore,

$$\mathsf{M}\left|\int \varphi(z)\,\mu_t^{(n,m)}(dz) - \int \varphi(z)\,\mu_t^{(n)}(dz)\right| \leqslant$$

$$\leqslant \mathsf{M}\frac{1}{n}\sum_{i=1}^{n} |\varphi(z_i^{(n)}(t)) - \varphi(z_i^{(n,m)}(t))| \leqslant$$

$$\leqslant \frac{K_1}{n}\sum_{i=1}^{n} \mathsf{M}\psi(z_i^{(n)}(t) - z_i^{(n,m)}(t)) = O(\varepsilon_m).$$

We shall assume, for the system (21), that for each m the conditions of Theorem 1 are fulfilled and that $\lambda_t^{(m)}(dz)$ is a limit statistical function. Then

$$\overline{\lim_{n\to\infty}}\, \mathsf{M}\left|\int \varphi(z)\,\mu_t^{(n)}(dz) - \int \varphi(z)\,\lambda_t^{(m)}(dz)\right| = O(\varepsilon_m),$$

$$\overline{\lim_{n\to\infty}}\, \mathsf{M}\left|\int \varphi(z)\,\mu_t^{(n)}(dz) - \int \varphi(z)\,\lambda_t^{(l)}(dz)\right| = O(\varepsilon_l)$$

and

$$\overline{\lim_{m,l\to\infty}}\left|\int \varphi(z)\,\lambda_t^{(m)}(dz) - \int \varphi(z)\,\lambda_t^{(l)}(dz)\right| = 0$$

for all $\varphi \in C_Z$ for which φ satisfies the Lipschitz condition. Using the inequality

$$\int \psi(z)\lambda_t^{(m)}(dz) \leqslant \overline{\lim_{n\to\infty}}\, \mathsf{M} \sum_{i=1}^{n} \psi(z_i^{(n,m)}(t)) \leqslant$$
$$\leqslant \overline{\lim_{n\to\infty}}\, \mathsf{M}\frac{1}{n}\sum_{i=1}^{n}\psi(z_i^{(n)}(t)) + \overline{\lim_{n\to\infty}}\, \mathsf{M}\frac{1}{n}\sum_{i=1}^{n}\psi(z_i^{(n)}(t) - z_i^{(n,m)}(t)) \leqslant c_1,$$

where c_1 is some constant, we convince ourselves that the set of measures $\{\lambda_t^{(m)}(dz)\}$ is compact. Therefore, for all $\varphi \in C_Z$,

$$\lim_{m\to\infty} \int \varphi(z)\lambda_t^{(m)}(dz) = \int \varphi(z)\lambda_t(dz),$$

where $\lambda_t(dz)$ is some normalized measure on Z. In this connection, for all $\varphi \in C_Z$,

$$\lim_{n\to\infty} \mathsf{M}\left|\int \varphi(z)\mu_t^{(n)}(dz) - \int \varphi(z)\lambda_t(dz)\right| = 0$$

(this relation is valid for φ satisfying the Lipschitz condition, and is fulfilled on the class of functions closed with respect to bounded convergence, uniform on each compactum).

Writing the relation (2) for the measure $\lambda_t^{(m)}$ and integrating it with respect to t, we obtain

$$\int \varphi(z)\lambda_t^{(m)}(dz) = \int \varphi(z)\lambda_0(dz) + \int_0^t \int\int \int_{\Theta_m} [\varphi(z) + f(\theta, z, z')) - \varphi(z)]\, m(d\theta)\,\lambda_s^{(m)}(dz)\,\lambda_s^{(m)}(dz')\,ds;$$

here, $\lambda_0(dz)$ is the limit measure for $\mu_0^{(n)}(dz)$. Using the existence and uniform boundedness of the integrals $\int\psi(z)\lambda_s^{(m)}(dz)$ and the estimate

$$\int_{\Theta\setminus\Theta_m} \psi(f(\theta, z, z'))\, m(d\theta) \leqslant \varepsilon_m(1 + \psi(z) + \psi(z')),$$

we convince ourselves that for $\varphi \in C_Z$ satisfying the Lipschitz condition the relation

$$\int \varphi(z)\lambda_t(dz) = \int \varphi(z)\lambda_0(dz) + \int_0^t \int\int\int_{\Theta} [\varphi(z) + f(\theta, z, z')) - \varphi(z)]\, m(d\theta)\,\lambda_s(dz)\,\lambda_s(dz')\,ds,$$

is valid and, differentiating it, we obtain (2). Thus, the following theorem is valid.

THEOREM 1'. Suppose the following conditions are satisfied:

(1) $Mp_{ij}^{(n)}(d\theta \times dt) = \frac{1}{n}m(d\theta)dt$, where m is a σ-finite measure,

(2) there exists a continuous function $\psi(z)$ coinciding with $|z|$ in a neighborhood of the point 0 and tending to ∞ as $|z| \to \infty$, such that relations (19) and (23) are fulfilled for some c, L,

(3) the sequence of measurable subsets $\Theta_n \subset \Theta$ increases, $\bigcup_n \Theta_n = \Theta$, for all n, $m(\Theta_n) < \infty$ and (22) is satisfied, and also the inequalities

$$\int_{\Theta_m} |f(\theta, z, z')| m(d\theta) \leqslant L_m(1 + |z| + |z'|),$$

$$\int_{\Theta_m} |f(\theta, z, z') - f(\theta, \bar{z}, \bar{z}')| m(d\theta) \leqslant L_m(|z - \bar{z}| + |z' - \bar{z}'|),$$

are satisfied for some L_m,

(4) the statistical distribution function for the solution (1'), $\mu_0^{(n)}(dz)$, converges to the limit measure $\lambda_0(dz)$ and $\overline{\lim_{n\to\infty}} \int\psi(z)\mu_0^{(n)}(dz) < \infty$.

Then, for all t > 0, the sequence of measures $\mu_t^{(n)}$ (of statistical distributions for solutions of equation (1') at the moment t) converges weakly with respect to a distribution to the nonrandom measure $\lambda_t(dz)$ such that for all $\varphi \in C_Z$ satisfying the Lipschitz condition, relation (2) is fulfilled.

REMARK 1. As was shown above, the solution of equation (2) under the conditions of Theorem 1 can be obtained as the weak limit of $\lambda_t^{(m)}(dz)$ of solutions of the equation

$$\frac{d}{dt}\int \varphi(z)\lambda_t^{(m)}(dz) = \int\int\int_{\Theta_m} [\varphi(z + f(\theta, z, z')) - \varphi(z)] \times \times m(d\theta)\lambda_t^{(m)}(dz)\lambda_t^{(m)}(dz')$$

with the initial condition $\lambda_0^{(m)}(dz) = \lambda_0(dz)$. If condition (3) of the theorem is fulfilled with the constants $L_m = L$ not depending on m, then the limit $\lambda_t^{(m)}(dz)$ exists for arbitrary sequence Θ_m for which $m(\Theta_m) < \infty$, $\Theta_m \subset \Theta_{m+1}$ and $\bigcup_m \Theta_m = \Theta$, where the limit does not depend on the choice of

Θ_m. In this sense the solution of equation (2) is unique.

REMARK 2. If the conditions of Theorem 1 are fulfilled, then the conditions of Theorem 1' are also fulfilled. To this end, one can put $\Theta_m = \Theta$ and take $\psi(z)$ to be the function $\psi(|z|)$, where $\psi(|z|)$ is the function introduced in the proof of Theorem 1 in subsection 2 (obviously, it can be chosen equal to $|z|$ in a neighborhood of zero). We now verify that conditions (19) and (23) are fulfilled. We have

$$\int_\Theta \psi(|f(\theta, z, z')|)\, m(d\theta) \leqslant m(\Theta)\psi\left(\frac{1}{m(\Theta)}\int_\Theta |f(\theta, z, z')|\, m(d\theta)\right) \leqslant$$
$$\leqslant m(\Theta)\psi\left[\frac{L}{m(\Theta)}(1+|z|+|z'|)\right) \leqslant c[1+\psi(|z|)+\psi(|z'|)]$$

for some c (we used the fact that $\psi(\lambda)$ is concave upward and subadditive). Analogously,

$$\int_\Theta \psi(|f(\theta, z, z')-f(\theta, \bar{z}, \bar{z}')|)\, m(d\theta) \leqslant$$
$$\leqslant m(\Theta)\psi\left(\frac{L}{m(\Theta)}(|z-\bar{z}|+|z'-\bar{z}'|)\right) \leqslant$$
$$\leqslant c_1[\psi(|z-\bar{z}|)+\psi(|z'-\bar{z}'|)].$$

Condition (22) is fulfilled trivially since $\Theta \setminus \Theta_m = \emptyset$ and hence the integral equals 0.

We now consider the limiting behavior of a separate particle under the assumption that the statistical distribution function of the system tends to a nonrandom limit. Inasmuch as the equations of the system are symmetric relative to individual particle, we shall consider the first particle:

$$dz_1^{(n)}(t) = \sum_{j=2}^{n} \int_\Theta f(\theta, z_1^{(n)}, z_j^{(n)})\, p_{1j}^{(n)}(d\theta \times dt).$$

We shall denote by $F_t^{(n)}$ the σ-algebra generated by the variables $z_i^{(n)}(s)$ for $s \leq t$ having values random measures $p_{ij}^{(n)}(d\theta \times ds)$ on $\Theta \times [0, t]$. Let $u(t, z)$ be a continuous bounded function on $[0, \infty) \times Z$ for which $\frac{\partial u}{\partial t}(t, z)$ is also continuous and bounded and for some L, $|u(t, z) - u(t, z_1)| \leq L|z - z_1|$. Using the Itô formula, we can write

$$u(t+h, z_1^{(n)}(t+h)) - u(t, z_1^{(n)}(t)) =$$
$$= \int_t^{t+h} \frac{\partial u}{\partial s}(s, z_1^{(n)}(s))\, ds + \sum_{j=2}^{n} \int_t^{t+h} \int_\Theta [u(s, z_1^{(n)}(s) +$$
$$+ f(\theta, z_1^{(n)}(s), z_j^{(n)}(s))) - u(s, z_1^{(n)}(s))]\, p_{1j}^{(n)}(d\theta \times ds).$$

We compute the conditional mathematical expectation relative to the σ-algebra $F_t^{(n)}$. We obtain

$$\mathsf{M}[u(t+h,\ z_1^{(n)}(t+h))/\mathcal{F}_t^{(n)}]=$$
$$=u(t,\ z_1^{(n)}(t))+\mathsf{M}\Big(\int_t^{t+h}\Big[\frac{\partial u}{\partial s}(s,\ z_1^{(n)}(s))+$$
$$+\frac{1}{n}\sum_{j=2}^{n}\int_\Theta[u(s,\ z_1^{(n)}(s)+f(\theta,\ z_1^{(n)}(s),\ z_j^{(n)}(s))-$$
$$-u(s,\ z_1^{(n)}(s))]\,m(d\theta)]\,ds/\mathcal{F}_t^{(n)}\Big)=$$
$$=u(t,\ z_1^{(n)}(t))+\mathsf{M}\Big(\int_t^{t+h}\Big[\frac{\partial u}{\partial s}(s,\ z_1^{(n)}(s))+$$
$$+\int\int_\Theta[u(s,\ z_1^{(n)}(s)+f(\theta,\ z_1^{(n)}(s),\ z))-$$
$$-u(s,\ z_1^{(n)}(s))\Big]\,m(d\theta)\,\mu_s^{(n)}(dz)\,ds/\mathcal{F}_t^{(n)}\Big).$$

We assume that the finite-dimensional distributions of the processes $z_1^{(m)}(t)$ converge to the finite-dimensional distributions of some random process $\tilde{z}(t)$. We set $F(s, z, z') = \int[u(s, z + f(\theta, z, z')) - u(s, z)]m(d\theta) + \frac{\partial}{\partial s} u(s, z)$. We shall assume that the conditions of Theorem 1' are fulfilled (it follows from Remark 2 that they are also fulfilled under the conditions of Theorem 1). Then $F(s, z, z')$ is a bounded continuous function. Therefore,

$$\sup_{\substack{|z-z_1|\leqslant\varepsilon\\ |z|\leqslant c}}\Big|\int F(s,\ z,\ z')\,\mu_s^{(n)}(dz')-\int F(s,\ z_1,\ z')\,\mu_s^{(n)}(dz')\Big|\leqslant$$
$$\leqslant 2\sup_{s,\ z'}|F(s,\ z,\ z')|\,\mu_s^{(n)}(\{z\colon\ |z|>r\})+$$
$$+\sup_{|z'|\leqslant r}\ \sup_{\substack{|z-z_1|\leqslant\varepsilon\\ |z|\leqslant c}}|F(s,\ z,\ z')-F(s,\ z_1,\ z')|,$$

that is, $\int F(s, z, z')\mu_s^{(n)}(dz')$ is uniformly continuous with respect to z on each bounded region uniformly relative to n. Using the convergence in probability of $\int F(s, z, z')\mu_s^{(n)}(dz')$ to $\int F(s, z, z')\lambda_s(dz')$, we obtain for all $c > 0$, that

$$\lim_{n\to\infty}\mathsf{M}\sup_{|z|\leqslant c}\Big|\int F(s,\ z,\ z')\,\mu_s^{(n)}(dz')-\int F(s,\ z,\ z')\,\lambda_s(dz')\Big|=0.$$

Consequently

$$\overline{\lim_{n\to\infty}} \mathsf{M} \left| \mathsf{M} \left(\int_t^{t+h} \left[\int F(s, z_1^{(n)}(s), z') \mu_s^{(n)}(dz') - \right.\right.\right.$$

$$\left.\left.\left. - \int F(s, z_1^{(n)}(s), z') \lambda_s(dz') \right] ds / \mathcal{F}_t^{(n)} \right) \right| \leqslant$$

$$\leqslant \overline{\lim_{n\to\infty}} \int_t^{t+h} \left[\mathsf{M} \sup_{|z| \leqslant r} \left| \int F(s, z, z') \mu_s^{(n)}(dz') - \right.\right.$$

$$\left.\left. - \int F(s, z, z') \lambda_s(dz') \right| + 2 \sup_{z, z'} |F(s, z, z')| \mathsf{P}\{|z_1^{(n)}(s)| > r\} \right] ds \leqslant$$

$$\leqslant 2 \int_t^{t+h} \sup_{z, z'} |F(s, z, z')| \mathsf{P}\{|z(s)| > r\} ds$$

and, inasmuch as the right member tends to zero as $r \to \infty$, then

$$\mathsf{M}(u(t+h, z_1^{(n)}(t+h) / \mathcal{F}_t^{(n)}) = u(t, z_1^{(n)}(t)) +$$
$$+ \mathsf{M}\left(\int_t^{t+h} F(s, z_1^{(n)}(s), z') \lambda_s(dz) ds / \mathcal{F}_t^{(n)} \right) + \alpha_n, \tag{24}$$

where $\lim_{n\to\infty} \alpha_n = 0$.

Suppose $t_1 < t_2 < \ldots < t_k \leqq t$ and that $g(z_1, \ldots, z_k)$ is a bounded continuous function of Z^k. We put

$$\hat{F}(s, z) = \int F(s, z, z') \lambda_s(dz').$$

This is a bounded continuous function of two variables. If, for all $\varepsilon > 0$,

$$\lim_{h\to 0} \overline{\lim_{n\to\infty}} \sup_{|s-s_1| \leqslant h} \mathsf{P}\{|z_1^{(n)}(s) - z_1^{(n)}(s_1)| > \varepsilon\} = 0, \tag{25}$$

then by the theorem on the convergence of distributions of integrals (see Gihman and Skorohod's <u>Introduction to the Theory of Random Processes</u> (Moscow: Nauka, 1965, p.628)), the joint distribution of the variables

$$z_1^{(n)}(t_1), \ldots, z_1^{(n)}(t_k), \quad \int_t^{t+h} \hat{F}(s, z_1^{(n)}(s)) ds$$

converges to the joint distribution of the variables

$$z_1(t), \ldots, z_1(t_k), \quad \int_t^{t+h} \hat{F}(s, z_1(s)) ds.$$

Multiplying (24) by $g(z_1^{(n)}(t_1), \ldots, z_1^{(n)}(t_k))$ and computing the unconditional mathematical expectation, and then passing to the limit, we find that

$$\begin{aligned} &\mathsf{M}g(z(t_1), \ldots, z(t_k))\, u(t+h, z(t+h)) = \\ &= \mathsf{M}g(z(t_1), \ldots, z(t_k)) \left[u(t, z(t)) + \int_t^{t+h} \hat{F}(s, z(s))\, ds \right]. \end{aligned} \tag{26}$$

We shall verify that condition (25) is also fulfilled under the conditions of Theorem 1'. Let $\psi(z)$ be the function given in the conditions of Theorem 1'. Then, using (19), we obtain

$$\begin{aligned} &\mathsf{M}\psi(z_1^{(n)}(t)) - \psi(z_1^{(n)}(0)) = \\ &= \frac{1}{n} \sum_{j=2}^{n} \mathsf{M} \int_0^t \int_\Theta [\psi(z_1^{(n)}(s) + f(\theta, z_1^{(n)}(s), z_j^{(n)}(s))) - \\ &\qquad - \psi(z_1^{(n)}(s))]\, m(d\theta)\, ds \leqslant \\ &\leqslant \frac{1}{n} \sum_{j=2}^{n} \mathsf{M} \int_0^t \int_\Theta \psi(f(\theta, z_1^{(n)}(s), z_j^{(n)}(s)))\, m(d\theta)\, ds \leqslant \\ &\leqslant \frac{c}{n} \sum_{j=2}^{n} \mathsf{M} \int_0^t (1 + \psi(z_1^{(n)}(s)) + \psi(z_j^{(n)}(s)))\, ds = \\ &= ct + c \int_0^t \mathsf{M}\psi(z_1^{(n)}(s))\, ds + c \int_0^t \mathsf{M} \int \psi(z)\, \mu_s^{(n)}(dz)\, ds. \end{aligned}$$

Using inequality (20), we convince ourselves that under the conditions of Theorem 1 we have

$$\sup_n \mathsf{M}\psi(z_1^{(n)}(t)) \leqslant \sup_n \psi(z_1^{(n)}(0)) + A(t),$$

where A(t) is an increasing function. Furthermore,

$$\begin{aligned} &\mathsf{M}\psi(z_1^{(n)}(t+h) - z_1^{(n)}(t)) = \\ &= \frac{1}{n} \mathsf{M} \sum_{j=2}^{n} \int_t^{t+h} \int_\Theta [\psi(z_1^{(n)}(s) + f(\theta, z_1^{(n)}(s), z_j^{(n)}(s)) - \\ &\qquad - z_1^{(n)}(t)) - \psi(z_1^{(n)}(s) - z_1^{(n)}(t))]\, m(d\theta)\, ds \leqslant \\ &\leqslant \frac{1}{n} \mathsf{M} \sum_{j=2}^{n} \int_t^{t+h} \int_\Theta \psi(f(\theta, z_1^{(n)}(s), z_j^{(n)}(s)))\, m(d\theta)\, ds \leqslant \\ &\leqslant c \int_t^{t+h} \mathsf{M}\left[1 + \psi(z_1^{(n)}(s)) + \int \psi(z)\, \mu_s^{(n)}(dz)\right] ds = O(h) \end{aligned}$$

uniformly relative to n provided $z_1^{(n)}(0)$ has a limit.

We denote by F_t the σ-algebra generated by the variables $\tilde{z}(s)$ for

$s \leq t$. Writing out the relations (26) and taking into account that g is arbitrary, we obtain

$$\mathsf{M}[u(t+h,\ \bar{z}(t+h))/\mathcal{F}_t] = \\ = u(t,\ \bar{z}(t)) + \mathsf{M}\Bigg(\int_t^{t+h}\Big[\frac{\partial}{\partial s}u(s,\ \bar{z}(s)) + \\ + \iint [u(s,\ \bar{z}(s) + f(\theta,\ \bar{z}(s),\ z)) - \\ - u(s,\ \bar{z}(s))]\, m(d\theta)\,\lambda_s(dz)\Big]ds/\mathcal{F}_t\Bigg).$$

We consider the Poisson measure with independent values $\hat{p}(d\theta \times dz \times ds)$ on $\Theta \times Z \times [0, \infty)$ for which

$$\mathsf{M}\hat{p}(d\theta \times dz \times ds) = m(d\theta)\,\lambda_s(dz)\,ds.$$

Let $\hat{z}(t)$ be the solution of the stochastic differential equation

$$d\hat{z}(t) = \int_\Theta \int_Z f(\theta,\ \hat{z}(t),\ z)\,\hat{p}(d\theta \times dz \times ds).$$

We shall assume that this equation has a unique solution on any interval of time with an arbitrary initial condition. We shall denote by $\hat{z}_{z,s}(t)$ the solution of equation (28) on the segment $[s, \infty)$ satisfying the initial condition $\hat{z}_{z,s}(s) = z$. Let

$$v(s,\ z) = \mathsf{M}\varphi(\hat{z}_{z,s}(t+h)).$$

If for all φ satisfying the Lipschitz condition the function v itself satisfies the Lipschitz condition, then it satisfies the equation

$$\frac{\partial}{\partial s}v(s,\ z) = \\ = -\iint [v(s,\ z + f(\theta,\ z,\ z')) - v(s,\ z))]\, m(d\theta)\,\lambda_s(dz') \tag{29}$$

(in particular, the derivative $\frac{\partial}{\partial s}v(s, z)$ exists and it is continuous and bounded). In fact, using the equations

$$\mathsf{M}v(s+\delta,\ \hat{z}_{z,s}(z+\delta)) = \\ = \mathsf{M}\varphi(\hat{z}_{s+\delta,\ \hat{z}_{z,s}(s+\delta)}(t+h)) = \mathsf{M}\varphi(\hat{z}_{z,s}(t+h)) = v(s,\ z),$$

that are valid for $s + \delta < t+h$, we shall have on the basis of Itô's formula that

$$\frac{1}{\delta}[v(s+\delta,\ z) - v(s+\delta,\ \hat{z}_{z,s}(s,\ \delta))] =$$

$$= -\frac{1}{\delta}\int_0^{\delta}\int\int [v(s+\delta,\ \hat{z}_{s,z}(s+\tau)+f(\theta,\ \hat{z}_{s,z}(s+\tau),\ z)) - \\ - v(s+\delta,\ \hat{z}_{s,z}(s+\tau))]\ \hat{p}(d\theta\times dz\times d\tau).$$

Computing the mathematical expectation and passing to the limit as $\delta \downarrow 0$ we obtain on the right the right member of (29) and on the left we shall have the right derivative of v with respect to s. Inasmuch as the right member of (29) is continuous with respect to s, from this it follows that the ordinary derivative also exists, which naturally coincides with the right derivative. This completes the proof of relation (29). We substitute v(s, z) into equation (27). Then by virtue of (29) the expression under the integral sign $\int_t^{t+h}$ vanishes. Taking into account that $u(t+h,\ z) = \varphi(z)$, we obtain

$$\mathsf{M}[\varphi(\tilde{z}(t+h))/\mathcal{F}_t] = u(t,\ \tilde{z}(t)).$$

Inasmuch as this equation is valid for all φ satisfying the Lipschitz condition, the conditional distribution $\tilde{z}(t+h)$ relative to the σ-algebra F_t depends only on $\tilde{z}(t)$. Therefore $\tilde{z}(t)$ will be a Markov process. Furthermore,

$$u(t,\ \tilde{z}(t)) = \mathsf{M}[\varphi(\hat{z}_{t,z}(t+h))]_{z=\tilde{z}(t)} = \\ = \int \hat{\mathsf{P}}(t,\ \tilde{z}(t),\ t+h,\ dz')\varphi(z'),$$

where $\hat{P}(t,\ z,\ t+h,\ dz')$ is the transition probability of the Markov process generated by solutions of equation (28). Hence

$$\mathsf{M}[\varphi(\tilde{z}(t+h))/\mathcal{F}_t] = \int \hat{\mathsf{P}}(t,\ \tilde{z}(t),\ t+h,\ dz')\varphi(z').$$

From this it follows that the transition probabilities of the processes $\hat{z}(t)$ and $\tilde{z}(t)$ coincide and hence their finite-dimensional distributions also coincide, provided the initial values coincide.

It follows from the boundedness in probability of the processes $z_1^{(n)}(t)$ under the condition that $z_1^{(n)}(0) = \tilde{z}(0)$ does not vary with n, that the sequence of processes $z_1^{(n)}(t)$ is compact with respect to distribution. On the other hand, under the indicated assumptions, every limiting process with respect to distributions will be a Markov process for some subsequence $z_1^{(n_1)}(t)$, being the solution of equation (28) with the initial condition $\hat{z}(0) = \tilde{z}(0)$ (to be precise, its distributions will coincide with the distribution of equation (28)). From this it follows that $z_1^{(n)}(t)$ will converge with respect to a distribution to the solution of equation (28) with the initial condition $\hat{z}(0) = \tilde{z}(0)$.

Finally, we shall establish the condition under which the function $v(s, z)$ satisfies the Lipschitz condition with respect to z if φ satisfies the Lipschitz condition. For $|\varphi(z) - \varphi(\bar{z})| \leq L z - \bar{z}|$, we have

$$|v(s, z) - v(s, \bar{z})| \leqslant L\mathsf{M}\,|\hat{z}_{s,z}(t+h) - \hat{z}_{s,\bar{z}}(t+h)|.$$

We shall assume that

$$\iint |f(\theta, z, z') - f(\theta, \bar{z}, z')|\, m(d\theta)\, \lambda_\tau(dz') \leqslant K\,|z - \bar{z}|. \tag{32}$$

Then

$$\begin{aligned}\mathsf{M}\,|\hat{z}_{s,z}(t) - \hat{z}_{s,\bar{z}}(t)| &\leqslant |z - \bar{z}| + \int_s^t \iint \mathsf{M}\,|f(\theta, \hat{z}_{s,z}(\tau), z') - \\ &\quad - f(\theta, \hat{z}_{s,\bar{z}}(\tau), z')|\, m(d\theta)\, \lambda_\tau(dz')\, d\tau \leqslant \\ &\leqslant |z - \bar{z}| + K \int_s^t \mathsf{M}\,|\hat{z}_{s,z}(\tau) - \hat{z}_{s,\bar{z}}(\tau)|\, d\tau.\end{aligned}$$

Hence,

$$\mathsf{M}\,|\hat{z}_{s,z}(t) - \hat{z}_{s,\bar{z}}(t)| \leqslant |z - \bar{z}|\, e^{K(t-s)},$$
$$|v(s, z) - v(s, \bar{z})| \leqslant L e^{K(t+h-s)}\, |z - \bar{z}|.$$

Now we can formulate the following theorem.

THEOREM 2. Suppose the conditions of Theorem 1' and condition (32) are fulfilled. If $z_1^{(n)}(0) = \tilde{z}(0)$, then the finite-dimensional distributions of sequences of processes $z_1^{(n)}(t)$ converge to the finite-dimension distributions of the process $\hat{z}(t)$, which is the solution of the stochastic differential equation (28) with initial condition $\hat{z}(0) = \tilde{z}(0)$.

It follows from the theorem just proved that, in particular, the motions of various particles are in the limit independent.

4. Limit Theorem for Statistical Distribution Functions. General Case

We consider the limiting behavior of the statistical distribution function for a system of equations of the form (1), §2. We shall need results on the compactness of sequences of statistical distribution functions for such systems and also some estimates of differences of integrals with respect to statistical distribution functions for two different systems.

LEMMA 1. Suppose that for the system (1), §2, one can indicate a function $\rho(t)$ that is defined, continuous and increasing to infinity on $[0, \infty)$, for which there exists a continuous derivative $\rho'(t)$, for some c, which satisfies the relation

$$\rho(t+h) \leqslant c[\rho(t)+\rho(h)]$$

and such that the coefficients of the system satisfy the following conditions:

(1) for some c_1 the inequalities

(a) $\frac{\rho'(|z|)}{|z|}(A(z) + a(z, z'), z) \leqq c_1[\rho(|z|) + \rho(|z'|)]$;

(b) $\int \rho(|f(\theta, z, z')|)m(d\theta) \leqq c_1[\rho(z) + \rho(|z'|)]$;

(2) $\overline{\lim_{n\ \infty}} \int \rho(|z|)\mu_0^{(n)}(dz) < \infty$.

are satisfied.

Then one can find an increasing function $c_1(t)$ such that for all t

$$\overline{\lim_{n\to\infty}} \mathsf{M} \int \rho(|z|)\mu_t^{(n)}(dz) \leqslant c_1(t).$$

Proof. Using Itô's formula, we can write

$$\rho(|z_i^{(n)}(t)|) + \rho(|z_i^{(n)}(0)|) + \int_0^t \frac{\rho'(|z_i^{(n)}(s)|)}{|z_i^{(n)}(s)|}(A(z_i^{(n)}(s)) +$$

$$+\frac{1}{n}\sum_{j=1}^n a(z_i^{(n)}(s),\ z_j^{(n)}(s),\ z_i^{(n)}(s))\,ds +$$

$$+\sum_{j=1}^n \int_0^t \int_\Theta [\rho(|z_i^{(n)}(s) + f(\theta,\ z_i^{(n)}(s),\ z_j^{(n)}(s))|) -$$

$$-\rho(|z_i^{(n)}(s)|)]\,m(d\theta)\,dS$$

(here we used the fact that the derivative in the direction u of the function $\rho(z)$ with respect to z equals

$$\rho'(|z|)\frac{(z,\ u)}{|z|}).$$

We compute the mathematical expectation in the preceding equation. Using (a) and (b) and also the inequalities

$$\frac{\rho'(|z_i|)}{|z_i|}\left(A(z_i)+\frac{1}{n}\sum_{j=1}^{n}a(z_i,\ z_j),\ z_i\right)=$$

$$=\frac{1}{n}\sum_{j=1}^{n}\frac{\rho(|z_i|)}{|z_i|}\left(A(z_i)+a(z_i,\ z_j),\ z_i\right)\leqslant$$

$$\leqslant\frac{c_1}{n}\sum_{j=1}^{n}[\rho(|z_i|)+\rho(|z_j|)]=c_1\left[\rho(|z_i|)+\frac{1}{n}\sum_{j=1}^{n}\rho(|z_j|)\right],$$

$$\int_\Theta[\rho(z_i+f(\theta,\ z_i,\ z_j))-\rho(z_i)]\,m(d\theta)\leqslant$$

$$\leqslant(1+c_1)m(\Theta)\rho(|z_i|)+\int_\Theta\rho(|f(\theta,\ z_i,\ z_j)|)\,m(d\theta)\leqslant$$

$$\leqslant[(1+c_1)m(\Theta)+c_1]\rho(|z_i|)+c_1\rho(|z_i|),$$

we convince ourselves that

$$\mathsf{M}\rho(|z_i^{(n)}(t)|)\leqslant\rho(|z_i^{(n)}(0)|)+$$

$$+((1+c_1)m(\Theta)+2c_1)\int_0^t\mathsf{M}\rho(|z_i^{(n)}(s)|)\,ds+$$

$$+\frac{2c_1}{n}\sum_{j=1}^{n}\int_0^t\mathsf{M}\rho(|z_j^{(n)}(s)|)\,ds.$$

Summing with respect to i and dividing by n, we obtain

$$\mathsf{M}\int\rho(|z|)\,\mu_t^{(n)}(dz)\leqslant\int\rho(|z|)\,\mu_0^{(n)}(dz)+$$

$$+((1+c_1)m(\Theta)+4c_1)\int_0^t\int\mathsf{M}\rho(|z|)\,\mu_s^{(n)}(dz)\,ds.$$

From this there follows the inequality

$$\varlimsup_{n\to\infty}\mathsf{M}\int\rho(|z|)\,\mu_t^{(n)}(dz)\leqslant$$

$$\leqslant\varlimsup_{n\to\infty}\int\rho(|z|)\,\mu_0^{(n)}(dz)\exp\{[(1+c_1)m(\Theta)+4c_1]t\},$$

which completes the proof of the lemma.

REMARK 1. If the measure $m(\Theta)$ can only be σ-finite, then instead of condition 1(b) one must place the condition 1(b'):

$$\int[\rho(|z+f(\theta,\ z,\ z')|)-\rho(|z|)]\,m(d\theta)\leqslant c_1(\rho(|z|)+\rho(|z'|)). \qquad (1)$$

Using this estimate we can, exactly as in the proof of the lemma, obtain the inequality

$$\overline{\lim_{n\to\infty}}\, \mathsf{M} \int \rho(|z|)\mu_\zeta^{(n)}(dz) \leqslant \overline{\lim_{n\to\infty}} \int \rho(|z|)\mu_0^{(n)}(dz)e^{A_{\Theta,1}t}. \tag{2}$$

If the sequence $\mu_o^{(n)}(dz)$ is compact, then we can find a sufficiently slowly increasing function $\rho(t)$ $(t \to \infty)$ such that

$$\overline{\lim_{n\to\infty}} \int \rho(|z|)\mu_0^{(n)}(dz) < \infty.$$

This function can be chosen to be concave upward. We shall look to see when condition (1) will be fulfilled for such $\rho(t)$. Suppose $A(|z|) \leqq c(1+|z|)$. Then

$$\frac{\rho'(|z|)}{|z|}(A(z),\ z) \leqslant c\rho'(|z|)[1+|z|] \leqslant c_1\rho(|z|),$$

since, for the indicated choice of $\rho(t)$, $\rho'(t)$ is bounded, $\rho'(|z|)|z| \leqq \rho(|z|) - \rho(0)$. In order to have

$$\frac{\rho'(|z|)}{|z|}(a(z,\ z'),\ z) \leqslant c_1(\rho(|z|)+\rho(|z'|)),$$

it suffices that the condition $|a(z, z')| \leqq c(1 + |z| + \rho(|z'|)$ is fulfilled. We note that inasmuch as $\int a(z_i^{(n)}(t), z')\mu_t^{(n)}(dz')$ occurs in the equation for $z_i^{(n)}(t)$, and this expression must be bounded as $n \to \infty$, then the assumption that $|a(z, z')|$ is estimated with respect to z' by the quantity $\rho(|z'|)$, is not very restrictive.

In order that condition 1(b) be fulfilled for $m(\Theta) < \infty$ it suffices that

$$\int f(\theta,\ z,\ z')m(d\theta) \leqslant c(1+|z|+|z'|).$$

In fact, by virtue of the fact that the function $\rho(t)$ is concave upward, we shall have

$$\int \rho(|f(\theta,\ z,\ z')|)m(d\theta) \leqslant m(\Theta)\rho\left(\frac{1}{m(\Theta)}\int |f(\theta,\ z,\ z')m(d\theta)\right) \leqslant$$
$$\leqslant m(\Theta)\rho\left(c\frac{|z|+|z'|+1}{m(\Theta)}\right).$$

Thus, if $A(z)$ and $\int f(\theta, z, z')m(d\theta)$ are linearly bounded at infinity (such boundedness is usually required for the coefficients of stochastic differential equations), then for the condition of the lemma to be fulfilled one must find a $\rho(t)$ that is concave upward so that condition (2) is fulfilled and for some c the inequality

$$|a(z,\ z')| \leqslant c(1+|z|+\rho(|z'|)).$$

holds.

In physical problems, the interaction forces depend on $|z-z'|$ in which connection variations of the velocities of randomly interacting particles do not surpass the sums of their velocities so that $|f(\theta, z, z')| \leq |z| + |z'|$ (recall that z is the pair (x; v) of a radius vector and a velocity so that $|v| \leq |z|$). Thus for such systems compactness will occur if A(z) is linearly bounded.

REMARK 2. Suppose $|A(z)| \leq c(1+|z|)$, $|a(z, z')| \leq c(1+|z|+\rho(|z'|)$ and condition (1) is fulfilled for some function $\rho(t)$, $t \in [0, \infty)$, that is concave upward and increases to $+\infty$. Then (2) is fulfilled in some c_1.

Suppose $\bar{z}_i^{(n)}$ is a solution of the system

$$dz_i(t) = \left(\bar{A}(z_i^{(n)}) + \frac{1}{n}\sum_{j=1}^{n} \bar{a}(z_i^{(n)}, z_j^{(n)})\right)dt + \\ + \sum_{j=1}^{n} \int \bar{f}(\theta, z_i^{(n)}, z_j^{(m)})\, p_{ij}^{(n)}(d\theta \times dt),$$

$\bar{z}_i^{(n)}(0) = z_i^{(n)}(0)$, $\bar{\mu}_t^{(n)}$ is a statistical distribution function for this system.

LEMMA 2. Suppose $\lambda(t)$ and $\rho(t)$ are two functions, defined on $[0, \infty)$, such that:

(a) for all $t \in [0, \infty)$, there exists $\lambda'(t)$, $\lambda'(t) > 0$ for $t > 0$, $\lambda(0) = 0$, $\lambda'(0) = 0$;

(b) for all $z, h \in Z$, the inequalities

$$\lambda(|z+h|) - \lambda(|z|) - \lambda'(|z|)\frac{(z, h)}{|z|} \leqslant c\lambda(|h|),$$

$$\lambda(|z+h|) \leqslant c[\lambda(|z|) + \lambda(|h|)],$$

are fulfilled for some c;

(c) $\rho(t) \to +\infty$ as $t \to \infty$.

We set

$$b(z, z') = A(z) + a(z, z') + \int f(\theta, z, z')\, m(d\theta),$$

$$\bar{b}(z, z') = \bar{A}(z) + \bar{a}(z, z') + \int \bar{f}(\theta, z, z')\, m(d\theta).$$

If the inequalities

$$(1)\quad (b(z, z') - b(\bar{z}, \bar{z}'), z - \bar{z})\lambda'(|z - \bar{z}|) \leqslant \\ \leqslant c_1 |z - \bar{z}|[\lambda(|z - \bar{z}|) + \lambda(|z' - \bar{z}'|)],$$

(2) $$\int \lambda(|f(\theta, z, z') - \bar{f}(\theta, \bar{z}, \bar{z}')|)\, m(d\theta) \leqslant \leqslant c_1 [\lambda(|z - \bar{z}|) + \lambda(|z' - \bar{z}'|)],$$

(3) $$\overline{\lim_{n\to\infty}}\, \mathsf{M} \int \rho(z)\, \mu_i^{(n)}(dz) \leqslant c(t),$$

where c(t) is some increasing function, are fulfilled for some c_1, then there exists an increasing $c_1(t)$ such that, for all n,

$$\mathsf{M}\frac{1}{n}\sum_{i=1}^{n} \lambda(|z_i^{(n)}(t) - \bar{z}_i^{(n)}(t)|) \leqslant \delta c_1(t),$$

where

$$\delta = \sup_{z, z'} \Big\{ |b(z, z') - \bar{b}(z, z')| + \int \lambda(|f(\theta, z, z') - \bar{f}(\theta, z, z')|)\, m(d\theta) \Big\} (\rho(z) + \rho(z'))^{-1}.$$

Proof. Applying Itô's formula to $\lambda(|z_i^{(n)}(t) - \bar{z}_i^{(n)}(t)|)$, we find that

$$\lambda(|z_i^{(n)}(t) - \bar{z}_i^{(n)}(t)|) = \int_0^t \frac{1}{n}\sum_{j=1}^{n} \frac{\lambda'(|z_i^{(n)}(s) - \bar{z}_i^{(n)}(s)|)}{|z_i^{(n)}(s) - \bar{z}_i^{(n)}(s)|} (b(z_i^{(n)}(s), z_j^{(n)}(s)) - \bar{b}(\bar{z}_i^{(n)}(s), \bar{z}_j^{(n)}(s), z_i^{(n)}(s) - \bar{z}_i^{(n)}(s)))\, ds +$$
$$+ \sum_{j=1}^{n} \int_0^t \int [\lambda(|z_i^{(n)}(s) - \bar{z}_i^{(n)}(s) + f(\theta, z_i^{(n)}(s), z_j^{(n)}(s)) - \bar{f}(\theta, z_i^{(n)}(s), z_j^{(n)}(s))|) - \lambda(|z_i^{(n)}(s) - \bar{z}_i^{(n)}(s)|) -$$
$$- \lambda'(|z_i^{(n)}(s) - \bar{z}_i^{(n)}(s)|) \Big(\frac{z_i^{(n)}(s) - \bar{z}_i^{(n)}(s)}{|z_i^{(n)}(s) - \bar{z}_i^{(n)}(s)|},$$
$$f(\theta, z_i^{(n)}(s), z_j^{(n)}(s)) - \bar{f}(\theta, z_i^{(n)}(s), z_j^{(n)}(s))\Big)]\, p_{ij}^{(n)}(d\theta \times ds).$$

We compute the mathematical expectation of this expression and then sum with respect to i. Making use of properties of the function $\lambda(t)$, we obtain

$$\mathsf{M}\frac{1}{n}\sum_{i=1}^{n} \lambda(|z_i^{(n)}(t) - \bar{z}_i^{(n)}(t)|) \leqslant$$
$$\leqslant c_1 \int_0^t \mathsf{M}\Big[\frac{1}{n}\sum_{i=1}^{n} \lambda(|z_i^{(n)}(s) - \bar{z}_i^{(n)}(s)|) + \frac{1}{n}\sum_{j=1}^{n} \lambda(|z_j^{(n)}(s) - \bar{z}_j^{(n)}(s)|)\Big] ds +$$

$$+\int_0^t \frac{1}{n^2}\sum_{i,j=1}^n \mathsf{M}\,|b(\bar z_i^{(n)}(s),\ \bar z_j^{(n)}(s)) - \bar b(z_i^{(n)}(s),\ z_j^{(n)}(s)|\,ds +$$

$$+c\int_0^t \frac{1}{n^2}\sum_{i,j=1}^n \int_\Theta \mathsf{M}\lambda(|f(\theta,\ z_i^{(n)}(s),\ z_j^{(n)}(s)) -$$

$$-\bar f(\theta,\ \bar z_i^{(n)}(s),\ \bar z_j^{(n)}(s))|)\,m(d\theta)\,ds \leqslant$$

$$\leqslant 2c_1(1+c^2)\int_0^t \mathsf{M}\frac{1}{n}\sum_{i=1}^n \lambda(|z_i^{(n)}(s) - \bar z_i^{(n)}(s)|)\,ds +$$

$$+\int_0^t \frac{1}{n^2}\sum_{i,j=1}^n \mathsf{M}(|b(\bar z_i^{(n)}(s),\ \bar z_j^{(n)}(s)) + \bar b(z_i^{(n)}(s),\ z_j^{(n)}(s))|) +$$

$$+c_2\int_\Theta \lambda(|f(\theta,\ z_i^{(n)}(s),\ z_j^{(n)}(s)) -$$

$$-\bar f(\theta,\ \bar z_i^{(n)}(s),\ \bar z_j^{(n)}(s))|)\,m(d\theta)\,ds) \leqslant$$

$$\leqslant c_2\left[\int_0^t \mathsf{M}\frac{1}{n}\sum_{i=1}^n \lambda(|z_i^{(n)}(s) - \bar z_i^{(n)}(s)|)\,ds + \right.$$

$$\left. +\delta\int_0^t \mathsf{M}\frac{1}{n}\sum_{i=1}^n \rho(|\bar z_i^{(n)}(s)|)\,ds\right],$$

where c_2 is some constant. Consequently

$$\mathsf{M}\frac{1}{n}\sum_{i=1}^n \lambda(|z_i^{(n)}(t) - \bar z_i^{(n)}(t)|) \leqslant c_2\delta\int_0^t \mathsf{M}\frac{1}{n}\sum_{i=1}^n \rho(|\bar z_i^{(n)}(s)|)\,ds\,e^{c_2 t}.$$

The proof of the lemma follows from this and condition (3).

REMARK. Functions of the form

$$\lambda(t) = \frac{t\,[(1+t)^\alpha - 1]}{1+t},\quad \lambda(t) = \frac{t\ln(1+t)}{1+t},$$

$$\lambda(t) = \frac{t^2}{1+t^2}.$$

can serve as examples of the function $\lambda(t)$ satisfying conditions (a) and (b). Condition (1) will be fulfilled if $|b(z, z')| \leqq c[\lambda(|z| + \lambda(|z'|)]$ and, moreover, the Lipschitz condition is satisfied.

THEOREM 1. Suppose that for the coefficients of the system (1), §2, the following conditions are fulfilled:

(1) for some $c > 0$,

$$|A(z) - A(\bar z)| + |a(z, z') - a(\bar z, \bar z')| +$$

$$+\int |f(\theta, z, z') - f(\theta, \bar z, \bar z')|\,m(d\theta) \leqslant c(|z - \bar z| + |z' - \bar z'|),$$

(2) there exist functions $\lambda(t)$ and $\rho(t)$ having the properties:

(a) for all $t \in [0, \infty)$ they are continuous, differentiable, $\lambda'(t) > 0$, $\rho'(t) > 0$, $\lambda(0) = 0$, $\rho(0) = 0$, $\lambda'(0) = 0$,

(b) for all $z, h \in Z$, the inequalities are fulfilled for some c,

$$\lambda(|z+h|)-\lambda(|z|)-\lambda'(|z|)\frac{(z, h)}{|z|}\leqslant c\lambda(|h|),$$
$$\lambda(|z+h|)\leqslant c[\lambda(|z|)+\lambda(|h|)],$$

(c) $\rho(t) \to \infty$ as $t \to \infty$, $\rho(t_1 + t_2) \leqq c[\rho(t_1) + \rho(t_2)]$ and the conditions (a) and (b) of Lemma 1 are fulfilled,

(d) $n\lambda\left(\frac{1}{n}|A(z)+a(z, z')|\right)\leqslant\varepsilon_n[1+\rho(|z|)+\rho(|z'|)]$,

where $\varepsilon_n \to 0$ as $n \to \infty$,

(e) $\overline{\lim\limits_{n\to\infty}}\int\rho(|z|)\mu_0^{(n)}(dz)<\infty$,

(f) $\lim\limits_{\gamma\to\infty}\sup\limits_{|z|+|z'|\geqslant\gamma}\dfrac{|A(z)|+|a(z, z')|+\int|f(\theta, z, z')|m(d\theta)}{\rho(|z|)+\rho(|z'|)}=0.$

Then for all t the sequnece of measures $\mu_t^{(n)}(dz)$ converges weakly with respect to a distribution to the non-random measure $\lambda_t(dz)$, moreover, for $\varphi \in C_Z$ the function $\int\varphi(z)\lambda_t(dz)$ is continuous in t, coincides with $\int\varphi(z)\lambda_o(dz)$ for $t = 0$ and, for all $\varphi \in C_Z^1$,

$$\frac{d}{dt}\int\varphi(z)\lambda_t(dz)=\int\left(\varphi'(z), A(z)+\int a(z, z')\lambda_t(dz')\right)\lambda_t(dz)+ \tag{3}$$
$$+\iiint[\varphi(z+f(\theta, z, z'))-\varphi(z)]m(d\theta)\lambda_t(dz)\lambda_t(dz').$$

Proof. We shall consider the measurable space $(\bar{\Theta}, \bar{\mathfrak{C}})$, where $\bar{\Theta}$ is obtained from Θ by adjoining a point $\bar{\theta}$ and the σ-algebra $\bar{\mathfrak{C}}$ consists of sets of the σ-algebra $\mathfrak{C}$ and sets of the form $C \cup \{\bar{\theta}\}$, where $C \in \mathfrak{C}$, $\{\bar{\theta}\}$ is a singleton containing $\bar{\theta}$. We set $\bar{m}(C) = m(C)$ for $c \in \mathfrak{C}$, $\bar{m}(C \cup \{\theta\}) = m(C) + \bar{m}$, where $C \in \mathfrak{C}$, $\bar{m} > 0$ is some natural number, and $\bar{m}(d\theta)$ is a measure on $\bar{\mathfrak{C}}$. Finally, we shall define a function $\bar{f}(\theta, z, z')$ for $\theta \in \bar{\Theta}$: $\bar{f}(\theta, z, z') = f(\theta, z, z')$ for $\theta \in \Theta$, $\bar{f}(\bar{\theta}, z, z') = \frac{1}{\bar{m}}[A(z) + a(z, z')]$. Suppose $\bar{z}_i^{(n)}(t)$ is a solution of the system of equations

$$dz_i^{(n)}(t)=\sum_{j=1}^{n}\int_{\Theta}\bar{f}(\theta, z_i^{(n)}, z_j^{(n)})p_{ij}^{(n)}(d\theta\times dt), \tag{4}$$

where $\bar{p}_{ij}^{(n)}(d\theta \times dt)$ is a Poisson measure with independent values on $\bar{\Theta} \times [0, \infty)$ for which $M\bar{p}_{ij}^{(n)}(d\theta \times dt) = \frac{1}{n}\bar{m}(d\theta)dt$. The initial conditions for $\bar{z}_i^{(n)}(t)$ are $\bar{z}_i^{(n)}(0) = z_i^{(n)}(0)$. We have

$$\int_{\bar{\Theta}} |\bar{f}(\theta, z, z')|\, m(d\theta) =$$
$$= \int_{\Theta} |f(\theta, z, z')|\, m(d\theta) + |a(z, z') + A(z)| \leqslant c(1 + |z| + |z'|),$$

$$\int_{\bar{\Theta}} |\bar{f}(\theta, z, z') - \bar{f}(\theta, \bar{z}, \bar{z}')|\, \bar{m}(d\theta) \leqslant$$
$$\leqslant \int_{\Theta} |f(\theta, z, z') - f(\theta, \bar{z}, \bar{z}')|\, m(d\theta) +$$
$$+ |a(z, z') - a(\bar{z}, \bar{z}')| + |A(z) - A(\bar{z})| \leqslant$$
$$\leqslant c(|z - \bar{z}| + |z' - \bar{z}'|).$$

Consequently, the conditions of Theorem 1, §3, are fulfilled for the system (4). For all $\varphi \in C_Z$ the limit

$$\lim_{n\to\infty} \int \varphi(z)\bar{\mu}_t^{(n)}(dz) = \int \varphi(z)\bar{\lambda}_t(dz)$$

exists in the sense of convergence in probability, and $\bar{\lambda}_t(dz)$ is a non-random probability measure for which

$$\frac{d}{dt}\int \varphi(z)\bar{\lambda}_t(dz) =$$
$$= \iiint [\varphi(z + \bar{f}(\theta, z, z')) - \varphi(z)]\, m(d\theta)\bar{\lambda}_t(dz)\bar{\lambda}_t(dz). \tag{5}$$

For the system (4), we have $\bar{A}(z) = 0$, $\bar{a}(z, z') = 0$ and

$$\bar{b}(z, z') = \int_{\bar{\Theta}} \bar{f}(\theta, z, z')\, \bar{m}(d\theta) =$$
$$= A(z) + a(z, z') + \int_{\Theta} f(\theta, z, z')\, m(d\theta) = b(z, z'),$$

whereas

$$\int \lambda(|f(\theta, z, z') - \bar{f}(\theta, z, z')|)\, \bar{m}(d\theta) =$$
$$= \bar{m}\lambda\left(\frac{|A(z) + a(z, z')|}{\bar{m}}\right) \leqslant \varepsilon_{\bar{m}}(\rho(|z|) + \rho(|z'|)).$$

By virtue of Lemma 2,

$$\mathsf{M}\frac{1}{n}\sum_{i=1}^{n}\lambda(|z_i^{(n)}(t)-\bar{z}_i^{(n)}(t)|)\leqslant\varepsilon_{\bar{m}}c_1(t), \tag{6}$$

where $c_1(t)$ is an increasing function that does not depend on n and $\bar{m}$.

It follows from Lemma 1 that

$$\mathsf{M}\frac{1}{n}\sum_{i=1}^{n}[\rho(|z_i^{(n)}(t)|)+\rho(|\bar{z}_i^{(n)}(t)|)]\leqslant\bar{c}_1(t), \tag{7}$$

where $\bar{c}_1(t)$ is an increasing function that does not depend on $\bar{m}$ and n. Let $\varphi \in C_Z$. For each $\varepsilon > 0$ we can find a c_ε such that

$$|\varphi(x)-\varphi(y)|\leqslant\varepsilon(1+\rho(x)+\rho(y))+c_\varepsilon\lambda(|x-y|).$$

Therefore,

$$\mathsf{M}\frac{1}{n}\sum_{i=1}^{n}|\varphi(z_i^{(n)}(t))-\varphi(\bar{z}_i^{(n)}(t))|\leqslant\varepsilon(1+2\bar{c}_1(t))+c_\varepsilon\varepsilon_{\bar{m}}c_1(t).$$

Consequently,

$$\overline{\lim_{n\to\infty}}\,\mathsf{M}\left|\int\varphi(z)\mu_t^{(n)}(dz)-\int\varphi(z)\bar{\lambda}_t(dz)\right|\leqslant \\ \leqslant\varepsilon(1+2\bar{c}_1(t))+c_\varepsilon\varepsilon_{\bar{m}}c_1(t). \tag{8}$$

Suppose $\tilde{m} > \bar{m}$ and $\tilde{\lambda}_t(dz)$ is the limiting statistical function for a system of the form (6) in which $\tilde{m}$ figures instead of $\bar{m}$. It follows from inequality (8) that

$$\left|\int\varphi(z)\bar{\lambda}_t(dz)-\int\varphi(z)\tilde{\lambda}_t(dz)\right|\leqslant2\varepsilon(1+2\bar{c}_1(t))+c_\varepsilon(\varepsilon_{\bar{m}}+\varepsilon_{\tilde{m}})c_1(t)$$

and hence the left member of this inequality tends to zero as $\bar{m}$, $\tilde{m}\to\infty$. This means that the measures $\bar{\lambda}_t(dz)$ converge weakly to some measure $\lambda_t(dz)$ as $\bar{m}\to\infty$. Passing to the limit as $\bar{m}\to\infty$ in (8), we obtain

$$\overline{\lim_{n\to\infty}}\,\mathsf{M}\left|\int\varphi(z)\mu_t^{(n)}(dz)-\int\varphi(z)\lambda_t(dz)\right|=0$$

(inasmuch as $\varepsilon > 0$ is arbitrary). From this follows the existence of the limiting distribution function for the sequence $\mu_t^{(n)}(dz)$. We denote $\bar{\lambda}_t(dz)$ obtained for the given choice of $\bar{m}$ by $\lambda_t^{\bar{m}}(dz)$. Integrating equation (5), and substituting there in advance the value $\bar{f}$ and the measure $\bar{m}(d\theta)$, we shall have

$$\int \varphi(z)\lambda_t^{\bar{m}}(dz) = \int \varphi(z)\lambda_0(dz) + \int_0^t \iiint [\varphi(z+f(\theta, z, z')) - \varphi(z)]\, m(d\theta)\,\lambda_s^{\bar{m}}(dz)\,\lambda_s^{\bar{m}}(dz')\,ds + \int_0^t \bar{m}\left[\varphi\left(z+\frac{1}{\bar{m}}(A(z)+a(z,z'))\right)-\varphi(z)\right]\lambda_s^{\bar{m}}(dz)\,\lambda_s^{\bar{m}}(dz')\,ds. \quad (9)$$

It follows from the weak convergence of $\lambda_t^{\bar{m}}$ to λ_t that the left member of (9) converges to $\int\varphi(z)\lambda_t(dz)$. Moreover, for all s, the functions

$$\iiint [\varphi(z+f(\theta, z, z')) - \varphi(z)]\, m(d\theta)\,\lambda_s^{\bar{m}}(dz)\,\lambda_s^{\bar{m}}(dz')$$

are bounded and converge to the function

$$\iiint [\varphi(z+f(\theta, z, z')) - \varphi(z)]\, m(d\theta)\,\lambda_s(dz)\,\lambda_s(dz')$$

($\int[\varphi(z + f(\theta, z, z')) - \varphi(z)]m(d\theta)$ is a continuous bounded function of two variables and $\lambda_s^{\bar{m}}(dz)\lambda_s^{\bar{m}}(dz')$ converges weakly in Z^2). Therefore, the second term in the right member of (9) converges to

$$\int_0^t \iiint [\varphi(z+f(\theta, z, z')) - \varphi(z)]\, m(d\theta)\,\lambda_s(dz)\,\lambda_s(dz')\,ds.$$

Now suppose $\varphi(z) \in C_Z^1$. Then the functions $\bar{m}[\varphi(z + \frac{1}{\bar{m}}(A(z) + a(z, z'))) - \varphi(z)]$ are bounded by the quantity $O(|A(z)| + |a(z, z')|)$ and converge to $(\varphi'(z), A(z) + a(z, z'))$ uniformly on each compact set since, by virtue of condition (e), for each $\delta > 0$ there exists a γ such that

$$\int_{|z|+|z'|>\gamma} (|A(z)|+|a(z, z')|)\,\lambda_s^{\bar{m}}(dz)\,\lambda_s^{\bar{m}}(dz') \leqslant \delta \int_{|z|+|z'|>\gamma} (\rho(|z|)+\rho(|z'|))\,\lambda_s^{\bar{m}}(dz)\,\lambda_s^{\bar{m}}(dz') \leqslant 2\delta \int \rho(z)\,\lambda_s^{\bar{m}}(dz) \leqslant 2\delta c_1(t)$$

on the basis of (8). For an arbitrary bounded continuous function $\Psi(t)$ that equals t for $|t| \leq \gamma$, we have

$$\lim_{\bar{m}\to\infty}\int_0^t\iint\psi\Big(\bar{m}\Big[\varphi\Big(z+\frac{1}{\bar{m}}(A(z)+$$
$$+a(z,z'))\Big)-\varphi(z)\Big]\Big)\lambda_s^{\bar{m}}(dz)\lambda_s^{\bar{m}}(dz')\,ds=$$
$$=\int_0^t\iint\psi((\varphi'(z),\ A(z)+a(z,\ z')))\lambda_s(dz)\lambda_s(dz')\,ds.$$

Using this equation and estimate (10) we convince ourselves that

$$\lim_{\bar{m}\to\infty}\int_0^t\iint\bar{m}\Big[\varphi\Big(z+\frac{1}{\bar{m}}(A(z)+a(z,\ z'))\Big)-$$
$$-\varphi(z)\Big]\lambda_s(dz)\lambda_s(dz')\,ds=$$
$$=\int_0^t(\varphi'(z),\ A(z)+a(z,\ z'))\lambda_s(dz)\lambda_s(dz')\,ds.$$

Passing to the limit as $\bar{m}\to\infty$ in (9), taking into account the preceding equations, we convince ourselves that equation (5) is valid. This completes the proof of the theorem.

REMARK 1. The theorem is also valid for the case when the measure $m(d\theta)$ is σ-finite. The only difference in the proof consists in that although $\int[\varphi(z + f(\theta, z, z')) - \varphi(z)]m(d\theta)$ will no longer be necessarily a bounded function, it is continuous; and for each $\delta > 0$ one can find a γ such that for $|z| + |z'| > \gamma$,

$$\Big|\int[\varphi(z+f(\theta,\ z,\ z'))-\varphi(z)]\,m(d\theta)\Big|\leqslant\delta(\rho(z)+\rho(|z'|))$$

for $\varphi\in C_Z^1$ by virtue of condition (e). Using estimate (7), from this we obtain that

$$\lim_{\bar{m}\to\infty}\int_0^t\iiint|\varphi(z+f(\theta,\ z,\ z'))-$$
$$-\varphi(z)|\,m(d\theta)\lambda_s^{\bar{m}}(dz)\lambda_s^{\bar{m}}(dz')\,ds=$$
$$=\int_0^t\iiint|\varphi(z+f(\theta,z,z'))-\varphi(z)|\,m(d\theta)\lambda_s(dz)\lambda_s(dz').$$

REMARK 2. If the functions $|A(z)|$, $|a(z, z')|$ and $\int|f(\theta, z, z')|m(d\theta)|$ are bounded, then condition (2) of the theorem is fulfilled if we take arbitrary $\lambda(t)$ and $\rho(t)$ for which (a), (b), (c), (f) are fulfilled (conditions (a), (b), (c) are not connected with the coefficients of the equation, and when $\mu_o^{(n)}$ is compact then $\rho(t)$ for which (f) is fulfilled always exists). In fact, $m\lambda(\frac{1}{m}|A(z) + a(z, z')|) \to \lambda'(0) = 0$ uniformly with respect to z, z' so that (d) holds for

$$\varepsilon_n = \sup_{z,z'} n\lambda(\frac{1}{n}|A(z) + a(z,z')|),$$

(d) follows from the boundedness of the numerator and the fact that the denominator tends to infinity.

We now consider the limiting behavior of a single particle.

THEOREM 2. Suppose the conditions of Theorem 1 are fulfilled and that $z_i^{(n)}(0)$ converges to $z_1(0)$. Then the finite-dimensional distributions of the process $z_1^{(n)}(t)$ converge to the finite-dimensional distribution of the process $\bar{z}(t)$, which is the solution of the stochastic differential equation

$$d\bar{z}(t) = \bar{a}(t, \bar{z}(t))\,dt + \int_Z \int_\Theta f(\theta, \bar{z}(t), z')\,\bar{p}(d\theta \times dz' \times dt)$$

with initial condition $z(0) = z_1(0)$, where $\bar{p}(d\theta \times dz \times dt)$ is a Poisson measure with independent values on $\Theta \times Z \times \{0, \infty)$ for which $M\bar{p}(d\theta \times dz \times dt) = m(d\theta)\lambda_t(dz)dt$, $a(t, z) = A(z) + \int a(z,z')\lambda_t(dz')$, and $\lambda_t(dz)$ is the limit statistical function of a system whose existence is asserted in Theorem 1.

Proof. Suppose $z_i^{(n,r)}(t)$ is the solution of the system of equations

$$dz_i^{(n,r)}(t) = \sum_{j=1}^{n} \int_\Theta f_r(\theta, z_i^{(n,r)}, z_j^{(n,r)})\,p_{ij}^{(n,r)}(d\theta \times dt), \tag{11}$$

where $z_i^{(n,r)}(0) = z_i(0)$, and the measure $p_{ij}^{(n,r)}$ and functions f_r are as in equation (4) if $\bar{m} = r$. It follows from Theorem 2, §3, that the finitedimensional distributions $z_1^{(n,r)}(t)$ converge to the finite-dimensional distributions of the process $\hat{z}_r(t)$ which is the solution of the stochastic differential equation

$$\begin{aligned} d\hat{z}_r(t) = \int_Z \frac{1}{r}[A(\hat{z}_r(t)) + a(\hat{z}_r(t), z')]\,\bar{p}_r(\{\bar{\theta}\} \times dz' \times dt) + \\ + \int_\Theta \int_Z f(\theta, \hat{z}_r(t), z')\,\bar{p}_r(d\theta \times dz' \times dt) \end{aligned} \tag{12}$$

with initial condition $z_1(0)$, the Poisson measure $\bar{p}_r(d\theta \times dz \times dt)$ has independent values on $\bar{\Theta} \times Z \times [0, \infty)$, $M\bar{p}_r(\{\bar{\theta}\} \times dz \times dt) = r\lambda_t^{(r)}(dz)dt$,

$M\bar{p}(d\theta \times dz \times dt) = m(d\theta)\lambda_t^{(r)}(dz)dt$ for $d\theta \quad \Theta$, $\lambda_t^{(r)}(dz)$ is the limiting statistical distribution function for the system (12). Using the estimates of Lemma 2, we obtain

$$\begin{aligned}
M\lambda(|z_1^{(n)}(t)-z_1^{(n,r)}(t)|) = \frac{1}{n}\sum_{j=1}^{n} M\int_0^t \Big\{ \int_\Theta [\lambda(|z_1^{(n)}(s) - z_1^{(n,r)}(s) + \\
+ f(\theta, z_1^{(n)}(s), z_j^{(n)}(s)) - f(\theta, z_1^{(n,r)}(s), z_j^{(n,r)})(s))|) - \\
- \lambda(|z_1^{(n)}(s) - z_1^{(n,r)}(s)|) - \\
- \frac{\lambda'(|z_1^{(n)}(s) - z_1^{(n,r)}(s)|)}{|z_1^{(n)}(s) - z_1^{(n,r)}(s)|}(f(\theta, z_1^{(n)}(s), z_j^{(n)}(s)) - \\
- f(\theta, z_1^{(n,r)}(s), z_j^{(n,r)}(s)), z_1^{(n)}(s) - z_1^{(n,r)}(s))]\, m(d\theta) + \\
+ r\Big[\lambda\Big(\Big| z_1^{(n)}(s) - z_1^{(n,r)}(s) + \frac{1}{r}(A(z_1^{(n)}(s)) + \\
+ a(z_1^{(n)}(s), z_j^{(n)}(s)))\Big|\Big) - \lambda(|z_1^{(n)}(s) - z_1^{(n,r)}(s)|) - \\
- \frac{\lambda'(|z_1^{(n)}(s) - z_1^{(n,r)}(s)|)}{r|z_1^{(n)}(s) - z_1^{(n,r)}(s)|}(A(z_1^{(n)}(s)) + \\
+ a(z_1^{(n)}(s), z_j^{(n)}(s)), z_1^{(n)}(s) - z_1^{(n,r)}(s))\Big\}\, ds.
\end{aligned}$$

Making use of a property of the function $\lambda(t)$, we obtain

$$\begin{aligned}
M\lambda(|z_1^{(n)}(t)-z_1^{(n,r)}(t)|) \leqslant cc_1\int_0^t M\lambda(|z_1^{(n)}(s) - z_1^{(n,r)}(s)|)\, ds + \\
+ M\frac{cc_1}{n}\sum_{j=1}^{n}\int_0^t \lambda(|z_j^{(n)}(s) - z_j^{(n,r)}(s)|)\, ds + \\
+ M\frac{c}{n}r\int_0^t \sum_{j=1}^{n}\lambda\Big(\frac{1}{r}|A(z_1^{(n)}(s)) + a(z_1^{(n)}(s), z_j^{(n)}(s))|\Big)\, ds \leqslant \\
\leqslant cc_1\int_0^t M\lambda(|z_1^{(n)}(s) - z_1^{(n,r)}(s)|)\, ds + cc_1\varepsilon_r c_1(t) + \\
+ c\varepsilon_r\int_0^t M\Big(\rho(|z_1^{(n)}(s)|) + \frac{1}{n}\sum_{j=1}^{n}\rho(|z_j^{(n)}(s)|)\Big)\, ds.
\end{aligned}$$

Under the conditions of Theorem 1,

$$\overline{\lim_{n\to\infty}}\, M\frac{1}{n}\sum_{j=1}^{n}\int_0^t \rho(|z_j^{(n)}(s)|)\, ds < \infty$$

(this follows from Lemma 1 and Remark 1 to it). Using the calculations made in Lemma 1, we can write the inequality

$$\mathsf{M}\rho(|z_1^{(n)}(t)|) \leqslant \rho(|z_1^{(n)}(0)|) + \\ + 2c_1\int_0^t \mathsf{M}\rho(|z_1^{(n)}(s)|)\,ds + 2c_1\mathsf{M}\frac{1}{n}\sum_{j=1}^{n}\int_0^t \rho(|z_j^{(n)}(s)|)\,ds.$$

Therefore,

$$\overline{\lim_{n\to\infty}}\, \mathsf{M}\rho(|z_1^{(n)}(t)|) \leqslant \\ \leqslant \left[\overline{\lim_{n\to\infty}}\, \rho(|z_1^{(n)}(0)|) + 2c_1 \overline{\lim_{n\to\infty}}\, \mathsf{M}\frac{1}{n}\sum_{j=1}^{n}\int_0^t \rho(|z_j^{(n)}(s)|)\,ds\right] e^{2c_1 t}.$$

There follows from this the existence of an increasing function $c_2(t)$ such that the inequality

$$\mathsf{M}\lambda(|z_1^{(n)}(t) - z_1^{(n,r)}(t)|) \leqslant \\ \leqslant cc_1\int_0^t \mathsf{M}\lambda(|z_1^{(n)}(s) - z_1^{(n,r)}(s)|)\,ds + \varepsilon_r c_2(t),$$

and hence also the inequality

$$\mathsf{M}\lambda(|z_1^{(n)}(t) - z_1^{(n,r)}(t)|) \leqslant \varepsilon_r c_2(t)\, e^{cc_1 t}.$$

is fulfilled. Passing to the limit in the inequality

$$\mathsf{M}\lambda(|z_1^{(n,r)}(t) - z_1^{(n,l)}(t)|) \leqslant c(\varepsilon_r + \varepsilon_l)\, c_2(t)\, e^{cc_1 t},$$

we obtain

$$\mathsf{M}\lambda(|\tilde{z}_l(t) - \tilde{z}_r(t)|) \leqslant c(\varepsilon_r + \varepsilon_l)\, c_2(t)\, e^{cc_1 t},$$

where the finite-dimensional distributions $\tilde{z}_1(t)$ coincide with the finite-dimensional distributions $\hat{z}_1(t)$. From this it follows that the finite-dimensional distributions of the processes $\hat{z}_r(t)$ converge as $r \to \infty$ to the finitedimensional distributions of some process. If we denote this process by $\hat{z}(t)$, then the finite-dimensional processes $z_1^{(n)}(t)$ will converge to its finite-dimensional distributions. In fact, if $\varphi_1, \ldots, \varphi_m$ are bounded functions that are uniformly continuous in the entire space, then, for $t_1 < \ldots < t_m$,

$$\left|\mathsf{M}\prod_{j=1}^{m}\varphi_j(z_1^{(n)}(t_j))-\mathsf{M}\prod_{j=1}^{m}\varphi_j(\hat{z}(t_j))\right|\leqslant\left|\mathsf{M}\prod_{j=1}^{m}\varphi_j(z_1^{(n)}(t_j))-\right.$$
$$\left.-\mathsf{M}\prod_{j=1}^{m}\varphi_j(z_1^{(n,r)}(t_j))\right|+\left|\mathsf{M}\prod_{j=1}^{m}\varphi_j(z_1^{(n,r)}(t_j))-\mathsf{M}\prod_{j=1}^{m}\varphi_j(\hat{z}_r(t_j))\right|+$$
$$+\left|\mathsf{M}\prod_{j=1}^{m}\varphi_j(\hat{z}_r(t_j))-\mathsf{M}\prod_{j=1}^{m}\varphi_j(\hat{z}(t_j))\right|.$$

The second term on the right tends to zero as $n \to \infty$, the first term tends to zero uniformly with repsect to n as $r \to \infty$, and the third term tends to zero as $r \to \infty$.

Let $\varphi \in C_Z^1$. Then, for arbitrary bounded continuous function $\Phi(z_1, \ldots, z_m)$ on Z^m and $t_1 < \ldots < t_m < t$ the equation

$$\begin{aligned}&\mathsf{M}\Phi(\hat{z}_r(t_1),\ldots,\hat{z}_r(t_m))[\varphi(\hat{z}_r(t))-\varphi(\hat{z}_r(t_m))]=\\&=\mathsf{M}\Phi(\hat{z}_r(t_1),\ldots,\hat{z}_r(t_m))\int_{t_m}^{t}\int_Z\Big(r\Big[\varphi\Big(\hat{z}_r(s)+\\&+\frac{1}{r}(A(\hat{z}_r(s))+a(\hat{z}_r(s),\ z))\Big)\Big)-\varphi(\hat{z}_r(s))\Big]+\\&+\int_\Theta[\varphi(\hat{z}_r(s)+f(\theta,\hat{z}_r(s),z))-\varphi(\hat{z}_r(s))]\Big)\lambda_s^{(r)}(dz)\,ds.\end{aligned}\tag{14}$$

holds. As $r \to \infty$, the function

$$\begin{aligned}\psi_r(s,\ z)=\int_Z\Big(r\Big[\varphi\Big(z+\frac{1}{r}(A(z)+a(z,\ z'))\Big)-\varphi(z)\Big]+\\+\int_\Theta[\varphi(z+f(\theta,\ z,\ z'))-\varphi(z)]\,m(d\theta)\,\lambda_s^{(r)}(dz')\end{aligned}$$

converges uniformly to the function

$$\begin{aligned}\psi(s,\ r)=(\varphi'(z),\ \bar{a}(s,\ z))+\\+\int_\Theta\int_Z[\varphi(z+f(\theta,\ z,\ z'))-\varphi(z)]\,m(d\theta)\,\lambda_s(dz').\end{aligned}\tag{15}$$

As $r \to \infty$, the left member of equation (14) converges to

$$\mathsf{M}\Phi(\hat{z}(t_1),\ \ldots,\ \hat{z}(t_m))[\varphi(\hat{z}(t))-\varphi(\hat{z}(t_m))].$$

If we prove that

$$\begin{aligned}&\lim_{r\to\infty}\mathsf{M}\Phi(\hat{z}_r(t_1),\ldots,\hat{z}_r(t_m))\int_{t_m}^{t}\psi_r(s,\ \hat{z}_r(s))\,ds=\\&=\mathsf{M}\Phi(\hat{z}(t_1),\ldots,\hat{z}(t_m))\int_{t_m}^{t}\psi(s,\ \hat{z}(s))\,ds,\end{aligned}\tag{16}$$

then we can pass to the limit in (14) and obtain the equation

$$\mathsf{M}\Phi(\hat z(t_1),\dots,\hat z(t_m))[\varphi(\hat z(t))-\varphi(\hat z(t_m))]=$$
$$=\mathsf{M}\Phi(\hat z(t_1),\dots,\hat z(t_m))\left\{\int_{t_m}^{t}(\varphi'(\hat z(s)),\ \bar a(s,\ \hat z(s)))\,ds+\right. \tag{17}$$
$$\left.+\int_{t_m}^{t}\int_{\Theta}[\varphi(\hat z(s)+f(\theta,\ \hat z(s),\ z'))-\varphi(\hat z(s))]\,m(d\theta)\lambda_s(dz')\,ds.\right.$$

In order to prove (16) it will suffice to show that the processes $\hat z_r(t)$ are uniformly statistically continuous relative to r. We have

$$\mathsf{M}\lambda(|\hat z_r(t+h)-\hat z_r(t)|)=$$
$$=\mathsf{M}\int_t^{t+h}\int_Z\Big[\Big|\lambda\Big(\Big|\hat z_r(s)+\frac{1}{r}(A(z)+a(z,\ z'))\Big|\Big)\Big|-$$
$$-\lambda(|\hat z_r(s)|)\Big]r+\int_{\Theta}[\lambda(|\hat z_r(s)+f(\theta,\ \hat z_r(s),\ z')|)-$$
$$-\lambda(|\hat z_r(s)|)]\,m(d\theta)\lambda_s^{(r)}(dz')\,ds\leqslant c_2\mathsf{M}\int_t^{t+h}\lambda(|\hat z_r(s)|)\,ds=O(h).$$

We made use here of a property of the function $\lambda(t)$ and of the uniform boundedness of $\mathsf{M}\lambda(\hat z_r(s))$, which follows from the preceding estimates. Equation (17) follows from this. Hence, $\hat z(t)$ is a Markov process which possesses the following property: for each function $\varphi \in C_Z^1$ the process

$$\varphi(\hat z(t))-\int_0^t\Big[(\varphi'(\hat z(s)),\ \bar a(s,\ \hat z(s)))+$$
$$+\int_{\Theta}\int_Z[\varphi(\hat z(s)+f(\theta,\ \hat z(s),\ z))-\varphi(\hat z(s))]\,m(d\theta)\lambda_s(dz)\Big]\,ds$$

is a martingale. It also follows from this that the finite-dimensional distributions of the process $\hat z(t)$ coincide with the finite-dimensional distributions of the process $\bar z(t)$ indicated in the formulation of the theorem. This completes the proof of the theorem.

We now consider the question of the convergence of distributions of functionals of the process $z_1^{(n)}(t)$. To this end, we need to investigate the compactness of distributions of processes of the type $z_1^{(n)}(t)$ in the space of functions without discontinuities of the second kind. We shall denote by $D_{[0,T]}(X)$, where X is some finite-dimensional linear space of function $x(t)$, $t \in [0, T]$, for which the right limits exist at each point $t \in [0, T)$, the left limits exist at each point $t \in (0, T]$, $x(t+) = x(t)$, $t < T$, $x(T) = x(T-)$. We introduce a metric ρ_D in $D_{[0,T]}(X)$ (see, for example, [15, Vol. 1, Ch. 6, §5]). The weak convergence and the compactness of measures in $D_{[0,T]}(X)$ are defined in the usual way. Let $\xi_n(t)$ be a sequence of random processes with trajectories in $D_{[0,T]}(X)$ if the sequence of measures μ_n corresponding to these processes is weakly compact in $D_{[0,T]}(X)$.

LEMMA 3. Suppose $\xi_n(t)$, $t \in [0, T]$ is a sequence of processes with values in Z, $F \subset C_Z$ a set of finite functions dense in the space of all finite functions from C_Z. A necessary and sufficient condition for the sequence of processes $\xi_n(t)$ in the space $D_{[0,T]}(Z)$ to be compact is that the followint two conditions be fulfilled:

(a) the sequence of quantities $\sup_{t \leq T} |\xi_n(t)|$ is bounded in probability;

(b) for each $\varphi \in F$ the sequence of processes $\varphi(\xi_n(t))$ is compact in $D_{0,T}(R)$.

Proof. The necessity of conditions (a) and (b) is obvious. To prove sufficiency we can consider only processes $\xi_n(t)$ that satisfy the condition $|\xi_n(t)| \leq c$ for some c since it follows from condition (a) that for sufficiently large T we can make

$$\sup_n \mathsf{P}\{\sup_{t \leqslant T} |\xi_n(t)| > c\}$$

as small as we wish. We shall now show that for each collection $\varphi_1, \ldots, \varphi_m \in F$ the sequences of processes $(\varphi_1(\xi_n(t)), \ldots, \varphi_m(\xi_n(t)))$ in R^m is compact in $D_{[0,T]}(R^m)$. We shall estimate the quantity by

$$\gamma_n(\varepsilon, h) = \mathsf{P}\Bigg\{\sup_{\substack{0 < s_1 < s_2 < s_3 \leqslant T \\ |s_3 - s_1| \leqslant h}} \min\Bigg[\sum_{k=1}^{m} |\varphi_k(\xi_n(s_1)) - \varphi_k(\xi_n(s_2))|^2; \sum_{k=1}^{m} |\varphi_k(\xi_n(s_2)) - \varphi_k(\xi_n(s_3))|^2\Bigg] > \varepsilon\Bigg\}.$$

For $\varphi, \psi \in F$, $h > 0$, we denote

$$\Delta_n(\varphi, \psi, h) = \sup_{\substack{0 \leqslant s_1 < s_2 < s_3 \leqslant T \\ |s_3 - s_1| \leqslant h}} \min[|\varphi(\xi_n(s_1)) - \varphi(\xi_n(s_2))|; |\psi(\xi_n(s_2)) - \psi(\xi_n(s_3))|].$$

Then

$$\gamma_n(\varepsilon, h) \leqslant \mathsf{P}\left\{\max_{i,j} \Delta_n(\varphi_i, \varphi_j, h) > \sqrt{\frac{\varepsilon}{m}}\right\},$$

since if

$$\sum_{k=1}^{m} (\varphi_k(\xi_n(s_1)) - \varphi_k(\xi_n(s_2)))^2 > \varepsilon$$

and

$$\sum_{k=1}^{m} (\varphi_k(\xi_n(s_2)) - \varphi_k(\xi_n(s_3)))^2 > \varepsilon,$$

then one can find i, j such that

$$|\varphi_i(\xi_n(s_1)) - \varphi_i(\xi_n(s_2))| > \sqrt{\frac{\varepsilon}{m}},$$

$$|\varphi_j(\xi_n(s_2)) - \varphi_j(\xi_n(s_3))| > \sqrt{\frac{\varepsilon}{m}}.$$

Therefore,

$$\gamma_n(\varepsilon, h) \leqslant \sum_{i,j=1}^{m} \mathsf{P}\left\{\Delta_n(\varphi_i, \varphi_j, h) > \sqrt{\frac{\varepsilon}{m}}\right\} \leqslant$$

$$\leqslant \sum_{i,j=1}^{m} \left[\mathsf{P}\left\{\Delta_n(\varphi_i + \varphi_j, \varphi_i + \varphi_j, h) > \frac{1}{2}\sqrt{\frac{\varepsilon}{m}}\right\} + \right.$$

$$\left. + \mathsf{P}\left\{\Delta_n(\varphi_i, \varphi_i, h) > \frac{1}{2}\sqrt{\frac{\varepsilon}{m}}\right\} + \mathsf{P}\left\{\Delta_n(\varphi_j, \varphi_j, h) > \frac{1}{2}\sqrt{\frac{\varepsilon}{m}}\right\}\right].$$

We used here the fact that if

$$|\varphi_i(\xi_n(s_1)) - \varphi_i(\xi_n(s_2))| > \sqrt{\frac{\varepsilon}{m}},$$

$$|\varphi_j(\xi_n(s_2)) - \varphi_j(\xi_n(s_3))| > \sqrt{\frac{\varepsilon}{m}},$$

$$\min\left[|\varphi_i(\xi_n(s_1)) - \varphi_i(\xi_n(s_2))|;\ |\varphi_i(\xi_n(s_2)) - \varphi_i(\xi_n(s_3))|\right] \leqslant$$

$$\leqslant \frac{1}{2}\sqrt{\frac{\varepsilon}{m}},$$

$$\min\left[|\varphi_j(\xi_n(s_1)| - \varphi_j(\xi_n(s_2))|;\ |\varphi_j(\xi_n(s_2)) - \varphi_j(\xi_n(s_3))|\right] \leqslant$$

$$\leqslant \frac{1}{2}\sqrt{\frac{\varepsilon}{m}},$$

then

$$|\varphi_i(\xi_n(s_1)) - \varphi_i(\xi_n(s_2))| - |\varphi_j(\xi_n(s_1)) - \varphi_j(\xi_n(s_2))| > \frac{1}{2}\sqrt{\frac{\varepsilon}{m}},$$

$$|\varphi_j(\xi_n(s_2)) - \varphi_j(\xi_n(s_3))| - |\varphi_i(\xi_n(s_2)) - \varphi_i(\xi_n(s_3))| > \frac{1}{2}\sqrt{\frac{\varepsilon}{m}},$$

and hence

$$|\psi_{i,j}(\xi_n(s_1)) - \psi_{i,j}(\xi_n(s_2))| > \frac{1}{2}\sqrt{\frac{\varepsilon}{m}}$$

and

$$|\psi_{i,j}(\xi_n(s_2)) - \psi_{i,j}(\xi_n(s_3))| > \frac{1}{2}\sqrt{\frac{\varepsilon}{m}},$$

where $\psi_{i,j} = \varphi_i + \varphi_j$. Therefore, $\sup_n \gamma_n(\varepsilon, h) \to 0$ for all $\varepsilon > 0$ as $h \to 0$. Moreover, it follows from condition (b) that for all $\varepsilon > 0$

$$\lim_{h\to 0} \sup_n \mathsf{P}\Big\{\sup_{s\leqslant h} \sum_{k=1}^{m} [(\varphi_k(\xi_n(s)) - \varphi_k(\xi_n(0)))^2 + \\ + (\varphi_k(\xi_n(T-s)) - \varphi_k(\xi_n(T)))^2] > \varepsilon\Big\} = 0.$$

It also follows from this that the sequence of processes $(\varphi_1(\xi_n(t)), \ldots, \varphi_m(\xi_n(t))$ is compact. If $X = R^m$ and $|\varphi_k(x) - x^k| \leq \delta$ for $|x| \leq c$ (x^k is the k-th coordinate of x), then

$$\sup_t |\xi_n(t) - \eta_n(t)| \leqslant \sqrt{m}\,\delta,$$

where $\eta_n(t)$ has a k-th coordinate $\varphi_k(\xi_n(t))$. The proof of the lemma now follows from the compactness of $\eta_n(t)$ in $D_{[0,T]}(R^{(m)})$ and the fact that $\delta > 0$ is arbitrary.

REMARK. If the sequence of processes $\xi_n(t)$ has the form $\xi_n(t) = \alpha_n(t) + \beta_n(t)$, where $\alpha_n(t)$ is a sequence of continuous processes and $\beta_n(t)$ is a sequence of processes in $D_{[0,T]}(X)$, where $\alpha_n(t)$ and $\beta_n(t)$ are compact, then the sequence $\xi_n(t)$ is also compact.

THEOREM 3. Under the conditions of Theorem 1, for arbitrary $T > 0$ the sequence of measures μ_n in $D_{[0,T]}(Z)$, corresponding to the processes $z_1^{(n)}(t)$, converges weakly to measure $\bar{\mu}$, corresponding to the process $\bar{z}(t)$ defined in Theorem 2.

Proof. It will suffice to verify the weak compactness of the sequence of processes $z_1^{(n)}(t)$ in $D_{[0,T]}(Z)$. We shall show that the $z_1^{(n)}(t)$ are uniformly bounded relative to n on each segment. Suppose $\rho_1(t)$ is a function satisfying the conditions: $\rho_1(t) \leq \rho(t)$; $|\rho_1'(t)| \leq 1$; $\rho_1(t) \to \infty$ as $t \to \infty$. Then

$$\rho_1(|z_1^{(n)}(t)|) \leqslant \rho_1(|z_1^{(n)}(0)|) + \frac{1}{n}\sum_{j=1}^{n}\int_0^t |\rho_1'(z_1^{(n)}(s))| \Big(\frac{z_1^{(n)}(s)}{|z_1^{(n)}(s)|}, A(z_1^{(n)}(s))\Big) + a(z_1^{(n)}(s), z_j^{(n)}(s))| ds + \frac{1}{n}\sum_{j=1}^{n}\int_0^t\int_\Theta |\rho_1(|z_1^{(n)}(s) + f(\theta, z_1^{(n)}(s), z_j^{(n)}(s))|) - \rho_1(|z_1^{(n)}(s)|)| p_{ij}^{(n)}(d\theta \times ds).$$

Since the right member of this inequality increases monotonely with t, we have

$$\sup_{t \leqslant T} \rho_1(|z_1^{(n)}(t)|) \leqslant \rho(|z_1^{(n)}(0)|) + \frac{1}{n}\sum_{j=1}^{n}\int_0^T |A(z_1^{(n)}(s)) + a(z_1^{(n)}(s), z_j^{(n)}(s))| ds + \sum_{j=1}^{n}\int_0^T\int_\Theta |f(\theta, z_1^{(n)}(s), z_j^{(n)}(s))| p_{ij}^{(n)}(d\theta \times ds).$$

Therefore, on the basis of conditions 2(e) of Theorem 1, for some c_2 we have

$$\mathbf{M}\sup_{t \leqslant T} \rho_1(|z_1^{(n)}(t)|) \leqslant \rho(|z_1^{(n)}(0)|) + c_2\Big[\mathbf{M}\int_0^T \rho(|z_1^{(n)}(s)|) ds + \frac{1}{n}\sum_{j=1}^{n}\mathbf{M}\int_0^T \rho(|z_j^{(n)}(s)|) ds\Big].$$

The right member of this inequality is bounded on the basis of inequalities (7) and (13). Thus, condition (a) of Lemma 3 is fulfilled. Writing the preceding inequality for z_i (instead of z_1) and summing on i, we obtain

$$\mathbf{M}\sup_{t \leqslant T}\frac{1}{n}\sum_{i=1}^{n}\rho_1(|z_i^{(n)}(t)|) \leqslant \frac{1}{n}\sum_{i=0}^{n}\rho(|z_i^{(n)}(0)|) + 2c_2\int_0^T \mathbf{M}\frac{1}{n}\sum_{i=1}^{n}\rho(|z_i^{(n)}(s)|) ds. \qquad (18)$$

It follows from this inequality that the values

$$\sup_{t\leq T}\frac{1}{n}\sum_{i=1}^{n}\rho_1(|z_i^{(n)}(t)|)$$

are bounded in probability

Furthermore, suppose $\varphi \in C_Z^1$ is finite. Then

$$\varphi(z_1^{(n)}(t))=\varphi(z_1^{(n)}(0))+\alpha_n(t)+\beta_n(t),$$

where

$$\alpha_n(t)=\frac{1}{n}\sum_{j=1}^{n}\int_0^t(\varphi'(z_1^{(n)}(s)),\ A(z_1^{(n)}(s))+a(z_1^{(n)}(s),\ z_j^{(n)}(s)))\,ds,$$

$$\beta_n(t)=\sum_{j=1}^{n}\int_0^t[\varphi(z_1^{(n)}(s)+f(\theta,\ z_1^{(n)}(s),\ z_j^{(n)}(s))-$$

$$-\varphi(z_1^{(n)}(s))]\,p_{1j}^{(n)}(d\theta\times ds).$$

By virtue of the remark to Lemma 3, it suffices to prove the compactness of $\alpha_n(t)$ separately and of $\beta_n(t)$ separately. It follows from the boundedenss and finiteness of $\varphi'(z)$ and condition 2(e) of Theorem 1 that for each $\varepsilon > 0$ one can find a c_ε such that $(\varphi'(z),\ A(z)\ +\ a(z,\ z'))\ \leq c_\varepsilon + \varepsilon\rho(|z'|)$. Therefore,

$$|\alpha_n(t+h)-\alpha_n(t)|\leqslant c_\varepsilon h+\varepsilon\int_t^{t+h}\frac{1}{n}\sum_{j=1}^{n}\rho(|z_j^{(n)}(s)|)\,ds\leqslant$$

$$\leqslant c_\varepsilon h+\varepsilon\int_0^T\frac{1}{n}\sum_{j=1}^{n}\rho(|z_j^{(n)}(s)|)\,ds.$$

The next inequality follows from this inequality:

$$\lim_{h\to 0}\sup_n \mathsf{P}\{\sup_{|s_1-s_2|\leqslant h}|\alpha_n(s_1)-\alpha_n(s_2)|>\delta\}\leqslant$$

$$\leqslant\frac{\varepsilon}{\delta}\sup_n \mathsf{M}\frac{1}{n}\sum_{j=1}^{n}\int_0^T\rho(|z_j^{(n)}(s)|)\,ds.$$

Since $\varepsilon > 0$ is arbitrary, the sequence $\alpha_n(t)$ is compact. Since, for some c',

$$|\beta_n(t+h)-\beta_n(t)|\leqslant$$

$$\leqslant c'\sum_{j=1}^{n}\int_t^{t+h}\int|f(\theta,\ z_1^{(n)}(s),\ z_j^{(n)}(s))|\,p_{1j}^{(n)}(d\theta\times ds),$$

it suffices to prove the compactness of the sequence of processes

$$\beta_n(t)=\sum_{j=1}^{n}\int_0^t\int|f(\theta,\ z_1^{(n)}(s),\ z_j^{(n)}(s))|\,p_{1j}^{(n)}(d\theta\times ds).$$

We introduce the random variables

$$\tau_{n,N}=\inf\left[t\leqslant T,\ \sup_{s\leqslant t}(\rho(z_1^{(n)}(s))+\frac{1}{n}\sum_{i=1}^{n}\rho(|z_i^{(n)}(s)|)>N\right];$$

if the set of such t is empty, then we set $\tau_{n,N} = T$. It follows from (18) and the boundedness of the right member of this inequality that with the choice of a sufficiently large N we can make $P\{\tau_{n,N} = T\}$ arbitrarily close to unity for all n. We set $\tilde{\beta}_n^{(N)}(t) = \bar{\beta}_n(\min[t,\ \tau_{n,N}])$. If the sequence $\tilde{\beta}_n^{(N)}(t)$ is compact for all N, then by virtue of the relation

$$\mathsf{P}\{\tilde{\beta}_n^{(N)}(t)=\beta_n(t),\ t\leqslant T\}=\mathsf{P}\{\tau_{n,N}=T\}$$

the compactness of the sequence $\bar{\beta}_n$ will follow from this. In order to prove the compactness of $\tilde{\beta}_n^{(N)}(t)$ we note that if the σ-algebra $F_t^{(n)}$ is generated by $p_{ij}^{(n)}(d\theta \times dt)$ for $s \leqq t$; i, $j \leqq n$, then, for $h > 0$,

$$\mathsf{M}[\tilde{\beta}_n^{(N)}(t+h)-\tilde{\beta}_n^{(N)}(t)/\mathscr{F}_t^{(n)}]=$$
$$=\mathsf{M}\left[\frac{1}{n}\sum_{j=1}^{n}\int_{\min[t,\tau_{n,N}]}^{\min[t+h,\tau_{n,N}]}\int_{\Theta}|f(\theta,\ z_1^{(n)}(s),\ z_j^{(n)}(s)|\,m(d\theta)\times\right.$$
$$\left.\times ds/\mathscr{F}_t^{(n)}\right]\leqslant c_1\int_{\min[t,\tau_{N,n}]}^{\min[t+h,\tau_{N,n}]}\mathsf{M}\left[\rho(|z_1^{(n)}(s)|)+\right.$$
$$\left.+\frac{1}{n}\sum_{j=1}^{n}\rho(|z_j^{(n)}(s)|)/\mathscr{F}_t^{(n)}\right]ds\leqslant c_1Nh$$

(here we made use of property 2(e) of Theorem 1). Therefore, for $t_1 < t_2 < t_3$,

$$\mathsf{M}|\tilde{\beta}_n^{(N)}(t_3)-\tilde{\beta}_n^{(N)}(t_2)|\cdot|\tilde{\beta}_n^{(N)}(t_2)-\tilde{\beta}_n^{(N)}(t_1)|\leqslant$$
$$\leqslant\mathsf{M}|\tilde{\beta}_n^{(N)}(t_2)-\tilde{\beta}_n^{(N)}(t_1)|\cdot\mathsf{M}[\tilde{\beta}_n^{(N)}(t_3)-\tilde{\beta}_n^{(N)}(t_2)/\mathscr{F}_{t_2}^{(n)}]\leqslant$$
$$\leqslant c_1^2N^2(t_3-t_2)(t_2-t_1)\leqslant c_1^2N^2(t_3-t_1)^2.$$

It follows from this estimate and Cencov's theorem (see [15, Vol. 1, p.508, Th. 3]) that the sequence $\tilde{\beta}_n^{(N)}(t)$ is compact and hence this completes the proof of the theorem.

REMARK. Under the conditions of Theorem 1, the sequence of processes $\eta_\varphi^{(n)}(t) = \int\varphi(z)\mu_t^{(n)}(dz)$ will also be weakly compact in $D_{[0,T]}(R)$ for $\varphi \in C_Z$, and therefore the quantity

$$\sup_{t\leqslant T}\left|\eta_\varphi^{(n)}(t) - \int \varphi(z)\lambda_t(dz)\right|$$

will converge to zero in probability. In order ot prove that the sequence of processes $\eta_\varphi^{(n)}(t)$ is compact, we note that, for $\varphi \in C_Z^1$,

$$|\eta_\varphi^{(n)}(t+h) - \eta_\varphi^{(n)}(t)| \leqslant \\ \leqslant c'\frac{1}{n}\sum_{i,j}\int_t^{t+h}\int_\Theta |f(\theta,\ z_i^{(n)}(s),\ z_j^{(n)}(s))|\, p_{ij}^{(n)}(d\theta\times ds),$$

for some c', so that it suffices to show the compactness of the sequence

$$\beta_n(t) = \frac{1}{n}\sum_{i,j=1}^n\int_0^t\int_\Theta |f(\theta,\ z_i^{(n)}(s),\ z_j^{(n)}(s))|\, p_{ij}(d\theta\times ds).$$

If we set

$$\hat{\tau}_{n,N} = \inf\left[t\leqslant T;\ \sup_{s\leqslant t}\frac{1}{n}\sum_{i=1}^n \rho(|z_i^{(n)}(s)|) > N\right],$$

then the compactness of $\hat{\beta}_n(\min[t,\ \hat{\tau}_{n,N}])$ is proved exactly as that of the sequence $\tilde{\beta}_n^{(N)}(t)$ in the proof of Theorem 3.

We shall now consider the joint limiting behavior of several particles. Without loss of generality, we may assume that these are the particles, 1, 2, ..., k. We write the following equations for them:

$$dz_i^{(n)}(t) = \left(A(z_i^{(n)}) + \frac{1}{n}\sum_{j=1}^n a(z_i^{(n)},\ z_j^{(n)})\right)dt + \\ + \sum_{j=1}^n\int_\Theta f(\theta,\ z_i^{(n)},\ z_j^{(n)})\, p_{ij}^{(n)}(d\theta\times dt).$$

The first two terms in the right member are asymptotically equivalent to the quantity $\bar{a}(t,\ z_i^{(n)}(t))$. We write the second sum on the right in the following form:

$$\int_{\Theta}\int_{Z} f(\theta,\ z_i^{(n)}(t),\ z)\, p_i^{(n)}(d\theta\times dz\times dt),$$

where

$$p_i^{(n)}(d\theta\times B\times dt)=\sum_{j=1}^{n}\chi_B(z_j^{(n)}(t))\, p_{ij}^{(n)}(d\theta\times dt).$$

As was shown in the proof of Theorem 2, $p_i^{(n)}(d\theta \times dz \times dt)$ converges weakly as $n \to \infty$ to the Poisson measure with the independent values $p_i(d\theta \times dz \times dt)$ for which

$$\mathsf{M}p_i(d\theta\times dz\times dt)=m(d\theta)\,\lambda_t(dz)\,dt.$$

We note now that, for $i \neq 1$,

$$\begin{aligned}\mathsf{M}(p_i^{(n)}(d\theta\times B\times dt)\,p_l^{(n)}(d\theta'\times B'\times dt)/\mathcal{F}_i^{(n)}) &= \\ =\frac{1}{n}\chi_B(z_i^{(n)}(t))\,\chi_B(z_j^{(n)}(t))\,m(d\theta)\,m(d\theta')\,dt &+ \\ +\mathsf{M}(p_i^{(n)}(d\theta\times B\times dt)/\mathcal{F}_i^{(n)}) &\times \\ \times\mathsf{M}(p_l^{(n)}(d\theta'\times B'\times dt)/\mathcal{F}_i^{(n)}) &+ O(1/n).\end{aligned}$$

It follows from this relation that the Poisson measures $p_i^{(n)}$ and $p_1^{(n)}$ are asymptotically independent and, moreover, that the measures $p_1^{(n)}, \ldots, p_k^{(n)}$ are simultaneously asymptotically independent. Therefore, it follows from the proof of Theorem 2 that under the condition that the $z_i^{(n)}(0)$, $i = 1, \ldots, k$, converge to $z_i(0)$, the joint distribution of the processes $\{z_i^{(n)}(t),\ i = 1, \ldots, k\}$ converges to the joint distribution of the processes $\{\bar z_i(t),\ i = 1, \ldots, k\}$, being the solution of the system of stochastic differential equations

$$\begin{aligned}d\bar z_i(t)&=\bar a(t,\ z_i(t))\,dt+\int f(\theta,\ \bar z_i(t),\ z)\,p_i(d\theta\times dz\times dt),\\ \bar z_i(0)&=z_i(0),\end{aligned}\tag{19}$$

where $p_i(d\theta \times dz \times dt)$ which are mutually independent Poisson measures for which

$$\mathsf{M}p_i(d\theta\times dz\times dt)=m(d\theta)\,\lambda_t(dz)\,dt.$$

Inasmuch as these measures are independent, the processes $\bar z_i(t)$ are also mutually independent. Making use of this situation and of the convergence in $D_{[0,T]}(z)$ of the distributions $z_i^{(n)}(t)$ to the distribution of the process $\bar z_i(t)$, we convince ourselves that the following theorem is valid.

THEOREM 4. Suppose the conditions of Theorem 1 are satisfied and that $z_i^{(n)}(0)$ converges to $z_i(0)$ as $n \to \infty$. Then the distribution of the k-dimensional process $(z_i^{(n)}(t), \ldots, z_k^{(n)}(t))$ converges weakly in $D_{[0,T]}(Z^k)$ to the distribution of the process $(\bar{z}_1(t), \ldots, \bar{z}_k(t))$, which is the solution of system (19) with the independent Poisson measures p_i.

COROLLARY. If $z_1^{(n)}(0), \ldots, z_k^{(n)}(0)$ are bounded, then for arbitrary bounded functions $G_1(z(\cdot)), \ldots, G_k(z(\cdot))$ that are continuous on $D_{[0,T]}(Z)$ in the metric ρ_D the equation

$$\lim_{n\to\infty}\left(M\prod_{i=1}^{k} G_i(z_i^{(n)}(\cdot)) - \prod_{i=1}^{k} MG_i(z_i^{(n)}(\cdot))\right) = 0.$$

holds.

Since it suffices to verify this relation on subsequences for which $\lim z_i^{(n)}(0)$, $i = 1, 2, \ldots, k$, exists, it follows from the independence of $\bar{z}_1(t), \ldots, \bar{z}_k(t)$.

5. Fluctuations

We now consider a system of the form given in the preceding section. Suppose $\mu_t^{(n)}(dz)$ is the statistical distribution function of the system and that $\lambda_t(dt)$ is the limit statistical distribution function. For sufficiently large n and sufficiently well-behaved sets A we have $\lim \mu_t^{(n)}(A) = \lambda_t(A)$. We shall be interested in the deviation $\mu_t^{(n)}(A) - \lambda_t(A)$ - the fluctuation of the number of particles in the phase volume A. For this purpose we shall study the limiting behavior of the sequence of (sign-alternating) random measures

$$\nu_t^{(n)}(A) = \sqrt{n}\,[\mu_t^{(n)}(A) - \lambda_t(A)].$$

The totality of these measures have unbounded variation (the variation of $\nu_t^{(n)}$ equals $2\sqrt{n}$). Therefore, one must not assume that one obtains a random measure in the limit. We corresesopnd to $\nu_t^{(n)}$ the linear random function defined on C_Z by

$$\nu_t^{(n)}(\varphi) = \int \varphi(z)\,\nu_t^{(n)}(dz),$$

and we shall study the limit distribution $\nu_t^{(n)}(\varphi)$; it is possible that it may not exist for all φ but rather only for φ from some linear subset $L \subset C_Z$. If the limit distribution of the quantities $\nu_t^{(n)}(\varphi)$ exists for all $\varphi \in L$ and it coincides with the distribution $\nu_t(\varphi)$, then for any finite collection $\varphi_1, \ldots, \varphi_m \subset L$ the joint distribution of the quantities $\nu_t^{(n)}(\varphi_1), \ldots, \nu_t^{(n)}(\varphi_m)$ converges to the joint distribution $\nu_t(\varphi_1), \ldots, \nu_t(\varphi_m)$. This follows from the fact that for all $\alpha_1, \ldots, \alpha_m \in R$, $\sum \alpha_k\varphi_k \in L$, and hence

$$\lim_{n\to\infty} \mathsf{M} \exp\left\{i \sum_{k=1}^{m} \alpha_k \nu_t^{(n)}(\varphi_k)\right\} = \lim_{n\to\infty} \mathsf{M} \exp\left\{i\nu_t^{(n)}\left(\sum_1^m \alpha_k\varphi_k\right)\right\};$$

exists; the continuity with respect to $\alpha_1, \ldots, \alpha_m$ of the right member follows from the compactness of the family of distributions of the vectors $\{(\nu_t^{(n)}(\varphi_1, \ldots, \nu_t^{(n)}(\varphi_m)), n = 1, 2, \ldots\}$.

We shall first stop to consider purely discontinuous processes. If $\varphi \in C_Z$, then on the basis of Itô's formula

$$\begin{aligned} d\int \varphi(z)\,\mu_t^{(n)}(dz) = \\ = \iiint [\varphi(z + f(\theta, z, z')) - \varphi(z)]\, m(d\theta)\, \mu_t^{(n)}(dz)\, \mu_t^{(n)}(dz')\, dt + \\ + \frac{1}{n} \sum_{i,j=1}^{n} \int [\varphi(z_i^{(n)} + f(\theta, z_i^{(n)}, z_j^{(n)})) - \varphi(z_i^{(n)})]\, q_{ij}^{(n)}(d\theta \times dt). \end{aligned}$$

We write the equation for the limit statistical distribution function (equation (2), §3):

$$\begin{aligned} d\int \varphi(z)\,\lambda_t(dz) = \\ = \iiint [\varphi(z + f(\theta, z, z')) - \varphi(z)]\, m(d\theta)\, \lambda_t(dz)\, \lambda_t(dz')\, dt. \end{aligned} \tag{1}$$

Subtracting this equation from the preceding one and multiplying the difference by $\sqrt{n}$ we obtain, after simple transformations, that

$$\begin{aligned} d\int \varphi(z)\,\nu_t^{(n)}(dz) = \\ = \iiint [\varphi(z + f(\theta, z, z')) - \varphi(z)]\, m(d\theta)\, \mu_t^{(n)}(dz)\, \nu_t^{(n)}(dz')\, dt + \\ + \iiint [\varphi(z + f(\theta, z, z')) - \\ - \varphi(z)]\, m(d\theta)\, \nu_t^{(n)}(dz)\, \lambda_t(dz')\, dt + d\eta_n(\varphi, t), \end{aligned} \tag{2}$$

where

$$d\eta_n(\varphi, t) = \frac{1}{\sqrt{n}} \sum_{1\le i<j\le n} \int_\Theta [\varphi(z_i^{(n)}(t) + f(\theta, z_i^{(n)}(t), z_j^{(n)}(t))) + \varphi(z_j^{(n)}(t)) + f(\theta, z_j^{(n)}(t), z_i^{(n)}(t)) - \varphi(z_i^{(n)}(t)) - \varphi(z_j^{(n)}(t))]\, q_{ij}^{(n)}(d\theta\times dt). \tag{3}$$

The process $\eta_n(\varphi, t)$ represents a martingale whose jumps do not exceed the quantity $4||\varphi||\frac{1}{\sqrt{n}}$. The characteristic of this martingale equals

$$\frac{1}{n^2}\sum_{1\le i<j\le n}\int_0^t\int_\Theta [\varphi(z_i^{(n)}(s) + f(\theta, z_i^{(n)}(s), z_j^{(n)}(s))) + \varphi(z_j^{(n)}(s) + f(\theta, z_j^{(n)}(s), z_i^{(n)}(s))) - \varphi(z_i^{(n)}(s)) - \varphi(z_j^{(n)}(s))]^2 m(d\theta)\, ds = \frac{1}{2}\int_0^t\iiint [Q_\theta\varphi(z, z')]^2 m(d\theta)\,\mu_s^{(n)}(dz)\,\mu_s^{(n)}(dz')\, ds,$$

where $Q_\theta\varphi(z, z') = \varphi(z + f(\theta, z, z')) + \varphi(z' + f(\theta, z, z')) - \varphi(z) - \varphi(z')$. As $n \to \infty$ the characteristic of the maringale $\eta_n(\varphi, t)$ converges to the nonrandom function of time

$$\frac{1}{2}\int_0^t\iiint [Q_\theta\varphi(z, z')]^2 m(d\theta)\,\lambda_s(dz)\,\lambda_s(dz')\, ds. \tag{4}$$

From this it follows that finite-dimensional distribtuions of a sequence of martingales $\eta_n(\varphi, t)$ are compact. If $\hat\eta(\varphi, t)$ is a process whose finite-dimensional distributions are limiting distributions for the finite-dimensional distributions $\eta_n(\varphi, t)$, then $\hat\eta(\varphi, t)$ will be a continuous martingale with nonrandom characteristic (4). Therefore, $\hat\eta(\varphi, t)$ is a Gaussian process with independent increments. Since the finite-dimensional distributions $\hat\eta(\varphi, t)$ are completely determined by the characteristic, we have that the sequence of finite-dimensional distributions $\eta_n(\varphi, t)$ has a unique limit point, and therefore this sequence converges. It has thus been established that finite-dimensional distributions of the processes $\eta_n(\varphi, t)$ converge to finite-dimensional distributions of a Gaussina process with independent increments whose mean equals zero and whose dispersion coincides with expression (4). Suppose $\gamma(d\theta \times dz \times dz' \times ds)$ is a Gaussian random measure with independent values on $\Theta \times Z^2 \times [0, \infty)$ for which

$$\mathsf{M}\gamma(d\theta\times dz\times dz'\times ds) = 0,$$
$$\mathsf{M}\gamma^2(d\theta\times dz\times dz'\times ds) = m(d\theta)\,\lambda_s(dz)\,\lambda_s(dz')\, ds.$$

Then, for each bounded measurable function $g(\theta, z, z', s)$ the random process

$$\zeta(t)=\int_0^t\int_\Theta\int_Z\int_Z g(\theta, z, z', s)\,\gamma(d\theta\times dz\times dz'\times ds)$$

will be a Guassian process with independent increments for which

$$M\zeta(t)=0,$$

$$M\zeta^2(t)=\int_0^t\int_\Theta\int_Z\int_Z g^2(\theta, z, z', s)\,m(d\theta)\,\lambda_s(dz)\,\lambda_s(dz')\,ds.$$

Therefore the finite-dimensional distribution of the process $\hat\eta(\varphi, t)$ coincides with the finite-dimensional distributions of the process

$$\zeta_\varphi(t)=\int_0^t\int\int\int Q_\theta\varphi(z, z')\,\gamma(d\theta\times dz\times dz'\times ds).$$

We assume that the $\nu_t^{(n)}(\varphi)$ have a limiting distribution coinciding with the distribution $\nu_t(\varphi)$. After passing to the limit in (1), we obtain the following equation for $\nu_t(\varphi)$:

$$d\nu_t(\varphi)=\int\int\int Q_\theta\varphi(z, z')\,m(d\theta)\,\nu_t(dz)\,\lambda_t(dz')\,dt+d\zeta_\varphi(t). \tag{5}$$

We introduce in C_Z the operator $S_t\varphi$ defined by the formula

$$S_t\varphi(z)=\int\int Q_\theta\varphi(z, z')\,m(d\theta)\,\lambda_t(dz'). \tag{6}$$

This is a bounded operator in C_Z. Equation (4) can be rewritten in the form

$$d\nu_t(\varphi)=\nu_t(S_t\varphi)\,dt+d\zeta_\varphi(t). \tag{7}$$

We shall show how equation (7) can be solved. Suppose that for $s < t$ the family of linear operators U_s^t in C_Z satisfies the relations

$$\frac{d}{d_s}U_s^t\varphi=-S_sU_s^t\varphi,\quad U_t^t\varphi=\varphi. \tag{8}$$

(The existence of such operators follows from the boundedness of S_s.) Then from (7) we obtain

$$d_s \nu_s (U_s^t \varphi) = \nu_s (dU_s^t \varphi) + d\nu_s (\psi)|_{\psi = U_s^t \varphi} =$$
$$= -\nu_s (S_s U_s^t \varphi)\, ds + \nu_s (S_s U_s^t \varphi)\, ds + d_s \zeta_\psi (s)|_{\psi = U_s^t \varphi} =$$
$$= \int\int\int Q_\theta U_s^t \varphi (z, z') \gamma (d\theta \times dz \times dz' \times ds).$$

Therefore,

$$\nu_t (\varphi) = \nu_0 (U_0^t \varphi) + \int_0^t \int\int\int Q_\theta U_s^t \varphi (z, z') \gamma (d\theta \times dz \times dz' \times ds). \qquad (9)$$

We shall derive formula (9) below.

THEOREM 1. Suppose the following conditions are satisfied: (1) the limiting statistical distribution function $\lambda_t(dz)$ exists and it satisfies equation (1); (2) for each $\varphi \in C_Z$ the limit

$$\nu_0 (\varphi) = \lim_{n \to \infty} \sqrt{n} \left[\int \varphi (z) \mu_0^{(n)} (dz) - \int \varphi (z) \lambda (dz) \right],$$

exist; (3) if $\hat{C}_Z$ is the set of those $\varphi \in C_Z$ for which $\lim_{z \to \infty} \varphi(z)$ exists, then for all $\varphi \in \hat{C}_Z$ we have

$$\int \varphi (z' + f (\theta, z', z))\, m (d\theta)\, \mu (dz') \in \hat{C}_Z.$$

for arbitrary finite measure $\mu(dz)$ on Z.

Then for each $\varphi \in C_Z$ the distribution $\nu_t^{(n)}(\varphi)$ converges to the distribution of the random variable $\nu_t(\varphi)$, defined by equation (9).

Proof. We denote by $S_t^{(n)}(\omega)$ that bounded random operator on C_Z defined by the equation

$$S_t^{(n)} (\omega) \varphi (z) = \int\int [\varphi (z + f (\theta, z, z')) - \varphi (z)]\, m (d\theta)\, \lambda_t (dz') +$$
$$+ \int\int [\varphi (z' + f (\theta, z', z)) - \varphi (z')]\, m (d\theta)\, \mu_t^{(n)} (dz'). \qquad (10)$$

With the aid of this operator, equation (2) can be written in the following form:

$$d\nu_t^{(n)} (\varphi) = \nu_t^{(n)} (S_t^{(n)} (\omega) \varphi)\, dt + d\eta_n (\varphi, t).$$

We define the family of random operators $U_s^t(n, \omega)\varphi$, $0 \leq s \leq t$, $\varphi \in C_Z$, $U_t^t(n, \omega)\varphi = \varphi$ for which

$$\frac{d}{ds} U_s^t(n, \omega)\varphi = -S_s^{(n)}(\omega) U_s^t(n, \omega)\varphi. \tag{11}$$

We denote

$$\mu_t^{(n)}(\varphi) = \int \varphi(z)\mu_t^{(n)}(dz).$$

Then

$$\mu_t^{(n)}(\varphi) - \mu_0^{(n)}(U_0^t(n, \omega)\varphi) =$$
$$= \sum_{i=1}^{\varkappa_n+1} [\mu_{\tau_i}^{(n)}(U_{\tau_i}^t(n, \omega)\varphi) - \mu_{\tau_{i-1}}^{(n)}(U_{\tau_{i-1}}^t(n, \omega)\varphi)],$$

where $0 < \tau_1 < \ldots < \tau_{k_n}$, are all the moments of the jumps of the n-dimensional process $(z_1^{(n)}(s), \ldots, z_n^{(n)}(s))$ on $[0, t]$, and $\tau_0 = 0$, $\tau_{k_{n+1}} = t$. Since the state of the n-dimensional process does not vary on the intervals (τ_{i-1}, τ_i), we have

$$\mu_{\tau_i}^{(n)} = \mu_{\tau_i+0}^{(n)} = \mu_{\tau_{i+1}-0}^{(n)}.$$

Consequently,

$$\mu_{\tau_i}^{(n)}(U_{\tau_i}^t(n, \omega)\varphi) - \mu_{\tau_{i-1}}^{(n)}(U_{\tau_{i-1}}^t(n, \omega)\varphi) =$$
$$= \mu_{\tau_i}^{(n)}(U_{\tau_i}^t(n, \omega)\varphi) - \mu_{\tau_i-0}^{(n)}(U_{\tau_i}^t(n, \omega)\varphi) +$$
$$+ \mu_{\tau_{i-1}}^{(n)}(U_{\tau_i}^t(n, \omega)\varphi - U_{\tau_{i-1}}^t(n, \omega)\varphi).$$

Moreover,

$$U_{\tau_i}^t(n, \omega)\varphi - U_{\tau_{i-1}}^t(n, \omega)\varphi = -\int_{\tau_{i-1}}^{\tau_i} S_u^{(n)}(\omega) U_u^t(n, \omega)\varphi\, du,$$

$$\mu_{\tau_{i-1}}^{(n)}\left(\int_{\tau_{i-1}}^{\tau_i} S_u^{(n)}(\omega) U_u^t(n, \omega)\varphi\, du\right) =$$
$$= \int_{\tau_{i-1}}^{\tau_i} \mu_u^{(n)}(S_u^{(n)}(\omega) U_u^t(n, \omega)\varphi)\, du.$$

Finally, we note that if the process $z_k^{(n)}$ and $z_j^{(n)}$ interchanged their states at the moment τ_i (i.e. an interaction between the k-th and j-th particles occurred) and $p_{kj}(\{\bar{\theta}\} \times \{\tau_i\}) = 1$, then

$$\mu_{\tau_i}^{(n)}(\varphi)-\mu_{\tau_i-0}^{(n)}(\varphi)=$$
$$=\frac{1}{n}[\varphi(z_k^{(n)}(\tau_i-0)+f(\bar{\theta},\ z_k^{(n)}(\tau_i-0),\ z_j^{(n)}(\tau_i-0)))-$$
$$-\varphi(z_k^{(n)}(\tau_i-0))+\varphi(z_j^{(n)}(\tau_i-0)+$$
$$+f(\bar{\theta},\ z_j^{(n)}(\tau_i-0),\ z_k^{(n)}(\tau_i-0)))-\varphi(z_j^{(n)}(\tau_i-0))]=$$
$$=\frac{1}{n}\sum_{k,\,j=1}^{n}\int_{\tau_{i-1}}^{\tau_i}[\varphi(z_k^{(n)}(s-0)+$$
$$+f(\theta,z_k^{(n)}(s-0),\ z_j^{(n)}(s-0)))-\varphi(z_k^{(n)}(s-0))]\,p_{kj}^{(n)}(d\theta\times ds).$$

Consequently,

$$\sum_{i=1}^{\varkappa_n}[\mu_{\tau_i}^{(n)}(U_{\tau_i}^t(n,\ \omega)\varphi)-\mu_{\tau_i-0}^{(n)}(U_{\tau_i}^t(n,\ \omega)\varphi)]=$$
$$=\frac{1}{n}\sum_{k,\,j=1}^{n}\int\int_0^t[U_s^t(n,\ \omega)\varphi(z_k^{(n)}(s-0)+$$
$$+f(\theta,\ z_k^{(n)}(s-0),\ z_j^{(n)}(s-0)))-$$
$$-U_s^t(n,\ \omega)\varphi(z_k^{(n)}(s-0))]\,p_{kj}^{(n)}(d\theta\times ds)$$

(the integral in the right member is taken as the usual integral for each ω; it is impossible to understand it to be a stochastic integral inasmuch the integral function depends on the future). Thus, the following equation is valid:

$$\begin{aligned}\mu_t^{(n)}(\varphi)-\mu_0^{(n)}(U_0^t(n,\ \omega)\varphi)=&\\=\frac{1}{n}\sum_{k,\,j=1}^{n}\int\int_0^t[U_s^t(n,\ \omega)\varphi(z_k^{(n)}(s-0)+&\\+f(\theta,\ z_k^{(n)}(s-0),\ z_j^{(n)}(s-0)))-&\\-U_s^t(n,\ \omega)\varphi(z_k^{(n)}(s-0))]\,p_{kj}^{(n)}(d\theta\times ds)-&\\-\int_0^t\mu_s^{(n)}(S_s^{(n)}(\omega)\,U_s^t(n,\ \omega)\varphi)\,ds.&\end{aligned}\tag{12}$$

Furthermore, we have

$$\frac{d}{ds}\int U_s^t(n,\ \omega)\varphi(z)\lambda_s(dz)=$$
$$=-\int S_s^{(n)}(\omega)\,U_s^t(n,\ \omega)\varphi(z)\lambda_s(dz)+$$
$$+\iiint[U_s^t(n,\ \omega)\varphi(z+f(\theta,\ z,\ z'))-$$
$$-U_s^t(n,\ \omega)\varphi(z)]\,m(d\theta)\lambda_s(dz)\lambda_s(dz)\lambda_s(dz')\,ds,$$

$$\int\varphi(z)\lambda_t(dz)-\int U_0^t(n,\ \omega)\varphi(z)\lambda_0(dz)=$$
$$=-\int_0^t\int S_s^{(n)}(\omega)\,U_s^t(n,\ \omega)\varphi(z)\lambda_s(dz)\,ds+$$

$$+\int_0^t \iiint [U_s^t(n,\omega)\varphi(z+f(\theta,z,z'))-$$
$$-U_s^t(n,\omega)\varphi(z)]\,m(d\theta)\,\lambda_s(dz)\,\lambda_s(dz')\,ds.$$

Subtracting the preceding equaiton from (12), multiplying the difference by $\sqrt{n}$ and then performing some straightforward transformations (taking the form of the operator $S_s^{(n)}(\omega)$ into account, we find that

$$\nu_t^{(n)}(\varphi)-\nu_0^{(n)}(U_0^t(n,\omega)\varphi)=$$
$$=\frac{1}{\sqrt{n}}\sum_{k,j=1}^{n}\int_0^t\int [U_s^t(n,\omega)\varphi(z_k^{(n)}(s-0)+$$
$$+f(\theta,z_k^{(n)}(s-0),z_j^{(n)}(s-0)))-$$
$$-U_s^t(n,\omega)\varphi(z_k^{(n)}(s-0))]\,q_{kj}^{(n)}(d\theta\times ds)=$$
$$=\frac{1}{\sqrt{n}}\sum_{k,j=1}^{n}\int_0^t [U_s^t\varphi(z_k^{(n)}(s)+f(\theta,z_k^{(n)}(s),z_j^{(n)}(s)))-$$
$$-U_s^t\varphi(z_k^{(n)}(s))]\,q_{kj}^{(n)}(d\theta\times ds)+\chi_\varphi^{(n)}(t),$$

where U_s^t is defined by formula (8) and

$$\chi_\varphi^{(n)}(t)=\frac{1}{\sqrt{n}}\sum_{k,j=1}^{n}\int_0^t\int \Psi_{k,j}^{(n)}(\varphi,\theta,s)\,q_{kj}^{(n)}(d\theta\times ds),$$
$$\Psi_{kj}^{(n)}(\varphi,\theta,s)=[U_s^t(n,\omega)-U_s^t]\varphi(z_k^{(n)}(s-0)+ \tag{13}$$
$$+f(\theta,z_k^{(n)}(s-0),z_j^{(n)}(s-0)))-$$
$$-[U_s^t(n,\omega)-U_s^t]\varphi(z_k^{(n)}(s-0)).$$

It follows from formula (10) that $||S_t^{(n)}(\omega)|| \leq 4m(\Theta)$. It follows from condition (3) of Theorem 1 that for each $\varphi \in \hat{C}_Z$ we have $S_t^{(n)}(\omega)\varphi \in \hat{C}_Z$ and

$$\|S_t^{(n)}(\omega)\varphi - S_t\varphi\| \to 0.$$

Using the equation

$$U_s^t(n,\omega)\varphi=\varphi+\int_s^t S_u^{(n)}(\omega)\,du+\dots$$
$$+\int\cdots\int_{s<u_1<\dots<u_k<t} S_{u_1}^{(n)}(\omega)\dots S_{u_k}^{(n)}(\omega)\varphi\,du_1\dots du_k+\dots$$

and the estimate

$$\left\| \int\limits_{s<u_1<\ldots<u_k<t}\!\!\cdots\!\!\int S_{u_1}^{(n)}(\omega)\ldots S_{u_k}^{(n)}(\omega)\varphi\, du_1\ldots du_k \right\| \leqslant$$
$$\leqslant \frac{(4m(\Theta))^k}{k!}(t-s)^k\|\varphi\|,$$

we can convince ourselves that

$$\|U_s^t(n,\omega)\varphi - U_s^t\varphi\| \to 0$$

for all $\varphi \in \hat{C}_Z$. Consequently, $\nu_0^{(n)}(U_0^t(n,\omega)\varphi) \to \nu_0(U_0^t\varphi)$. Repeating the argument which helped us to find the form of the limiting distribution for the process $\eta_n(\varphi, t)$ defined by equation (3), we convince ourselves that the limiting distribution of the process

$$\frac{1}{\sqrt{n}}\sum_{k,j=1}^{n}\int\limits_0^t [U_s^t\varphi(z_k^{(n)}(s)+f(\theta, z_k^{(n)}(s), z_j^{(n)}(s))) -$$
$$- U_s^t\varphi(z_k^{(n)}(s))]\, q_{kj}^{(n)}(d\theta\times ds)$$

coincides with the distribution of the process

$$\int\limits_0^t\iiint Q_\theta U_s^t\varphi(z,z')\,\gamma(d\theta\times dz\times dz'\times ds).$$

Therefore, to prove the theorem it suffices to show that $\chi_\varphi^{(n)}(t) \to 0$ in probability. We have

$$\mathsf{M}[\chi_\varphi^{(n)}(t)]^2 =$$
$$= \frac{1}{n}\sum_{k,j,l,i=1}^{n}\int\limits_0^t\int\limits_0^t\iint \mathsf{M}\Psi_{kj}^{(n)}(\varphi,\theta_1,s_1)\,\Psi_{li}^{(n)}(\varphi,\theta_2,s_2)\times$$
$$\times q_{kj}^{(n)}(d\theta_1\times ds_1)\, q_{li}^{(n)}(d\theta_2\times ds_2) +$$
$$+ \frac{1}{n}\sum_{k,j=1}^{n}\int\limits_0^t\int\limits_\Theta \mathsf{M}[\Psi_{kj}^{(n)}(\varphi,\theta,s)]^2[q_{kj}^{(n)}(d\theta\times ds)]^2.$$

In order to calculate the mathematical expectations in the right member of the preceding equation, we need some auxiliary constructions. We shall consider $F_t^{(n)}$-measurable random variables, where $F_t^{(n)}$ is the σ-algebra generated by the measure $p_{ij}^{(n)}(d\theta \times ds)$, $i,j \leqq n$ for $s \leqq t$. Let ξ be such a random variable. We denote by $\xi_{ij}(d\theta \times ds)$ the value of this random variable under the condition that $p_{ij}^{(n)}(d\theta \times ds) = r$, $r = 0,1$. Then

$$\xi - [\xi_{ij}^0(d\theta\times ds) + (\xi_{ij}^1(d\theta\times ds) - \xi_{ij}^0(d\theta\times ds))\, p_{ij}^{(n)}(d\theta\times ds)]$$

is different from zero if, and only if, $p_{ij}^{(n)}(d\theta \times ds) > 1$, and the probability of this event is $O(n^{-1}m(d\theta)ds)^2$.

We note that ξ_{ij}^0 and ξ_{kj}^1 doe not depend in $p_{ij}^{(n)}(d\theta \times ds)$. We shall be interested in the random variables ξ expressed in terms of $z_k^{(n)}(s)$, $s \leqq t$. We shall show how $\xi_{ij}^0(d\theta \times ds)$ and $\xi_{ij}^1(d\theta \times ds)$ are calculated in case. To this end, we shall first construct the process

$$z_k^{(n)}(d\theta \times ds,\ u,\ 0,\ i,\ j), \quad z_k^{(n)}(d\theta \times ds,\ u,\ 1,\ i,\ j).$$

Here $z_k^{(n)}(B, u, 0, i, j)$, where B is a measurable set in $\Theta \times [0, \infty)$, is the solution of (1), §3, if in the solution the Poisson measure $p_{ij}^{(n)}(d\theta \times dt)$ is replaceable by the Poisson measure $p_{ij}^{(n)}(d\theta \times dt \setminus B)$. For $B = C \times [t_1, t_2]$, $z_k^{(n)}(B, u, 1, i, j)$ satisfies precisely such an equation on $[0, t_1]$ and (t_1, ∞), and, moreover,

$$z_i^{(n)}(t_1) - z_i^{(n)}(t_1 - 0) = f(\bar{\theta},\ z_i^{(n)}(t_1 - 0),\ z_j^{(n)}(t_1 - 0)),$$
$$z_j^{(n)}(t_1) - z_j^{(n)}(t_1 - 0) = f(\bar{\theta},\ z_j^{(n)}(t_1 - 0),\ z_i^{(n)}(t_1 - 0)),$$

where $\bar{\theta} \in C$. If $\xi = F(z_1^{(n)}(\cdot), \ldots, z_n^{(n)}(\cdot))$ is some measurable function of $z_1^{(n)}(\cdot), \ldots, z_n^{(n)}(\cdot)$, then in order to calculate $\xi_{ij}^0(d\theta \times ds)$ one must take the processes $z_k^{(n)}(d\theta \times ds, \cdot, 0, i, j)$ as the arguments of the funciton F and the processes $z_k^{(n)}(d\theta \times ds, 1, i, j)$ in order to calculate $\xi_{ij}^1(d\theta \times ds)$.

Let $s_1 < s_2$, $C_n \in \Theta$ and $\cap C_n = \{\bar{\theta}\}$. Then

$$M\xi_{ij}^0(C_n \times [s_1, s_2]) \to M\xi_{ij}^0(s, \bar{\theta}),$$

whereas

$$M\xi_{ij}^1(C_n \times [s_1, s_2]) \to M\xi_{ij}^1(s, \bar{\theta})$$

when

$$s_1 \to s,\ s_2 \to s,\ n \to \infty,$$

where $\xi_{ij}^1(s, \bar{\theta})$ is the result of substituting the processes $z_k^{(n)}(\bar{\theta}, s, u, 1, i, j)$ which are obtained from $z_k^{(n)}(C \times [s_1, s_2], u, 1, i, j)$

for $C = \{\bar{\theta}\}$, $s_1 = s_2 = s$, in the expression for F. We note that the processes $z_k^{(n)}(\bar{\theta}, s, u, 1, i, j)$ satisfy with respect to u for $u \neq s$ the same system of equations as does $z_k^{(n)}(u)$, and in the point $u = s_1$, we have

$$z_k^{(n)}(s-0) =$$

$$= z_k^{(n)}(s),\ k \neq i, j;\ z_i^{(n)}(s) - z_i^{(n)}(s-0) = f(\bar{\theta},\ z_i^{(n)}(s-0), z_j^{(n)}(s-0)), z_j^{(n)}(s) - z_j^{(n)}(s-0) = f(\bar{\theta}, z_j^{(n)}(s-0), z_j^{(n)}(s-0)).$$

In order to calculate

$$M\xi\,[q_{ij}^{(n)}(d\theta \times ds)]^2$$

we make use of the independence of ξ_{ij}^0 and ξ_{ij}^1 from $q_{ij}^{(n)}(d\theta \times ds)$. We shall have

$$\begin{aligned} M\xi\,[q_{ij}^{(n)}(d\theta\times ds)]^2 &= M\xi_{ij}^0(d\theta\times ds)\,m(d\theta)\,ds + \\ &+ M[\xi_{ij}^1(d\theta\times ds) - \xi_{ij}^0(d\theta\times ds)]M[q_{ij}^{(n)}(d\theta\times ds)]^2\,p_{ij}^{(n)}(d\theta\times ds) + \\ &+ o\left(\frac{m(d\theta)\,ds}{n}\right) = \frac{1}{n}M\xi_{ij}^0(d\theta\times ds)\,m(d\theta)\,ds + \\ &+ \frac{1}{n}M[\xi_{ij}^1(d\theta\times ds) - \xi_{ij}^0(d\theta\times ds)]\,m(d\theta)\,ds + \\ &+ o\left(\frac{m(d\theta)\,ds}{n}\right) = \frac{1}{n}M\xi_{ij}^1(d\theta\times ds)\,m(d\theta)\,ds + o\left(\frac{m(d\theta)\,ds}{n}\right). \end{aligned}$$

We denote by $\hat{\Psi}_{ij}^{(n)}(\varphi, \theta_1, s_1, d\theta \times ds)$ the magnitude $\xi_{ij}^1(d\theta \times ds)$ calculated for $\xi = [\Psi_{ij}^{(n)}(\varphi, \theta_1, s_1)]^2$. Then

$$\int_0^t \int_\Theta M[\Psi_{ij}^{(n)}(\varphi, \theta, s)]^2 [q_{ij}(d\theta\times ds)]^2 =$$

$$= \frac{1}{n}\int_0^t \int_\Theta M\hat{\Psi}_{ij}^{(n)}(\varphi, \theta, s, \theta, s)\,m(d\theta)\,ds.$$

In calculating $\hat{\Psi}_{ij}^{(n)}(\varphi, \theta, s, \theta, s)$ it will be necessary to substitute the function $z_k^{(n)}(\theta, s, u, 1, i, j)$ in the expression for $U_s^t(n, \omega)$. We denote the result of this substitution by $\hat{U}_s^t(n, \omega)$, and the result of such a substitution in the operator $S_u^n(\omega)$ by $\hat{S}_u^n(\omega)$. We note that in order to calculate the operator $\hat{S}_u^n(\omega)$ one must substitute in the right member of expression (10), instead of $\mu_u^{(n)}(dz')$, the statistical distribution function $\hat{\mu}_u^n(dz')$ for $\{z_k^{(n)}(\theta, s, u, 1, i, j), k = 1, \ldots,$

n}. If $z_k^{(n)}(\theta, s, u, 1, i, j) \neq z_k^{(n)}(u)$ $(s < u < t)$, then either $k = i$ or $k = j$, or the k-th particle interacted at some moment $u_1 \in [u, s]$ with the l-th particle, for which $z_l^{(n)}(u_1 - 0) \neq z_l^{(n)}(\theta, s, u_1 - 0, 1, i, j)$. We denote by $\rho_n(u)$ the number of particles for which $z_k^{(n)}(u_1)$ and $z_k^{(n)}(\theta, s, u_1, 1, i, j)$ do not coincide on $[s, u]$. Then,

$$\begin{gathered}\mathsf{P}\{\rho_n(u+\Delta u)-\rho_n(u)>1\}=o(\Delta u),\\ \mathsf{P}\{\rho_n(u+\Delta u)-\rho_n(u)=1/\rho_n(u)=m\}=\\ =\mathsf{P}\Big\{\sum_{\substack{1\leqslant l\leqslant m\\ r>m}} p_{lr}^{(n)}(\Theta\times\Delta u)=1\Big\}=\frac{m(n-m)}{n}m(\Theta)\,\Delta u+o(\Delta u)\end{gathered} \tag{14}$$

(there must transpire an interaction of one of the m fixed particles with one of the remaining n-m particles).

It follows from relations (14) that

$$\begin{aligned}\mathsf{M}(\rho_n^2(u+\Delta u)/\rho_n(u))={}&\\ ={}&\rho_n^2(u)+2\rho_n(u)\,\mathsf{M}\,[\rho_n(u+\Delta u)-\rho_n(u)/\rho_n(u)]+\\ &+\mathsf{M}([\rho_n(u+\Delta u)-\rho_n(u)]^2/\rho_n(u))\leqslant\\ \leqslant{}&\rho_n^2(u)+2\rho_n(u)\frac{\rho_n(u)(n-\rho_n(u))}{n}m(\Theta)\,\Delta u+\\ &+\frac{\rho_n(u)(n-\rho_n(u))}{n}m(\Theta)\,\Delta u+o(\Delta u).\end{aligned}$$

Therefore,

$$\begin{gathered}\frac{d}{du}\mathsf{M}\rho_n^2(u)\leqslant[3\mathsf{M}\rho_n^2(u)+\mathsf{M}\rho_n(u)]\,m(\Theta)\leqslant\\ \leqslant\frac{7}{2}m(\Theta)\,\mathsf{M}\rho_n^2(u)+\frac{1}{2}m(\Theta),\\ \frac{d}{du}\Big(\mathsf{M}\rho_n^2(u)+\frac{1}{7}\Big)\leqslant\frac{7}{2}m(\Theta)\Big(\mathsf{M}\rho_n^2(u)+\frac{1}{7}\Big),\\ \mathsf{M}\rho_n^2(u)\leqslant\Big(4+\frac{1}{7}\Big)\exp\Big\{\frac{7}{2}m(\Theta)(u-s)\Big\}-\frac{1}{7}\end{gathered}$$

(here we used the fact that $\rho(s) = 2$). It is easy to see that

$$\Big|\int\varphi(z)\,\mu_n^{(n)}(dz)-\int\varphi(z)\,\hat{\mu}_n^{(n)}(dz)\Big|\leqslant\frac{\|\varphi\|}{n}\rho_n(u)\leqslant\frac{\|\varphi\|}{n}\rho_n(t).$$

It follows from this inequality that

$$\|S_n^{(n)}(\omega)-\hat{S}_n^{(n)}(\omega)\|\leqslant\frac{2m(\Theta)}{n}\rho_n(t).$$

Moreover,

$$\|S_n^{(n)}(\omega)\|\leqslant 4m(\Theta),\quad \|\hat{S}_n^{(n)}(\omega)\|\leqslant 4m(\Theta)$$

and hence

$$\|U_u^t(n,\ \omega)\| \leqslant e^{4m(\Theta)(t-u)}, \quad \|\hat{U}_u^t(\omega)\| \leqslant e^{4m(\Theta)(t-u)}.$$

Thus,

$$\|U_u^t(n,\ \omega) - \hat{U}_u^t(n,\omega)\| \leqslant \int_u^t \|S_\tau^{(n)}(\omega)\, U_\tau^t(n,\ \omega) -$$

$$- \hat{S}_\tau^{(n)}(\omega)\hat{U}_\tau^t(n,\omega)\|\, d\tau \leqslant \int_u^t \|S_\tau^{(n)}(\omega) - \hat{S}_\tau^{(n)}(\omega)\| \cdot \|U_\tau^t(n,\omega)\| d\tau +$$

$$+ \int_u^t \|\hat{S}_\tau^{(n)}(\omega)\| \cdot \|U_\tau^t(n,\ \omega) - \hat{U}_\tau^t(n,\ \omega)\|\, d\tau \leqslant$$

$$\leqslant \frac{c_1}{n}\rho_n(t) + c_1 \int_u^t \|U_\tau^t(n,\ \omega) - \hat{U}_\tau^t(n,\ \omega)\|\, d\tau,$$

where c_1 is some constant. From the preceding inequality we find that

$$\|U_u^t(n,\ \omega) - \hat{U}_u^t(n,\ \omega)\| \leqslant \frac{c_1}{n}\rho_n(t)\, e^{c_1(t-u)}. \tag{15}$$

Consequently, there exists a constant c_2 such that

$$\mathsf{M}\Psi_{ij}^{(n)}(\varphi,\ \theta,\ s,\ \theta,\ s) \leqslant \frac{c_2}{n} + \mathsf{M}[\Psi_{ij}^{(n)}(\varphi,\ \theta,\ s)]^2,$$

and hence,

$$\int_0^t \int_\Theta \mathsf{M}[\Psi_{ij}^{(n)}(\varphi,\ \theta,\ s)]^2 [q_{ij}^{(n)}(d\theta \times ds)]^2 \leqslant \frac{c_2}{n^2} m(\Theta)\, t +$$

$$+ \frac{1}{n}\int_0^t \int_\Theta \mathsf{M}[\Psi_{ij}^{(n)}(\varphi,\ \theta,\ s)]^2 m(d\theta)\, ds.$$

Thus,

$$\frac{1}{n}\sum_{k,j=1}^{n} \int_0^t \int_\Theta \mathsf{M}[\Psi_{kj}^{(n)}(\varphi,\ \theta,\ s)]^2 [q_{ij}^{(n)}(d\theta \times ds)]^2 =$$

$$= O\left(\frac{1}{n}\right) + \int_0^t \int_\Theta \frac{1}{n^2} \sum_{k,j=1}^{n} \mathsf{M}[\Psi_{kj}^{(n)}(\varphi,\ \theta,\ s)]^2 m(d\theta)\, ds =$$

$$= O\left(\frac{1}{n}\right) + \mathsf{M}\int_0^t \int_\Theta \int\int \{[(U_s^t(n,\ \omega) - U_s^t)\varphi](z + f(\theta,\ z,\ z')) -$$

$$- [(U_s^t(n,\ \omega) - U_s^t)\varphi](z)\}^2 \mu_{s-0}^{(n)}(dz)\, \mu_{s-0}^{(n)}(dz')\, m(d\theta)\, ds.$$

Using the boundedness of the integrand function and the fact that

$$\| U_s^t(n, \omega)\varphi - U_s^t\varphi \| \to 0,$$

we convince ourselves that

$$\lim_{n\to\infty} \frac{1}{n} \sum_{k,j=1}^{n} \int_0^t \int_\Theta \mathsf{M}[\Psi_{kj}^{(n)}(\varphi, \theta, s)]^2 [q_{kj}^{(n)}(d\theta \times ds)]^2 = 0.$$

Calculation of the quantity

$$\mathsf{M}\xi q_{kj}^{(n)}(d\theta_1 \times ds_1)\, q_{li}^{(n)}(d\theta_2 \times ds_2)$$

is carried out analogously to that of

$$\mathsf{M}\xi [q_{kj}^{(n)}(d\theta \times ds)]^2.$$

We introduce the quantities $\xi(k, j, l, i, 0, 0)$, $\xi(k, j, l, i, 1, 0)$, $\xi(k, j, l, i, 0, 1)$, $\xi(k, j, l, i, 1, 1)$, which are calculated in the propositions

(1) $p_{kj}^{(n)}(d\theta_1 \times ds_1) = 0, \quad p_{li}^{(n)}(d\theta \times ds_2) = 0;$

(2) $p_{kj}^{(n)}(d\theta_1 \times ds_1) = 1, \quad p_{li}^{(n)}(d\theta_2 \times ds_2) = 0;$

(3) $p_{kj}^{(n)}(d\theta_1 \times ds_1) = 0, \quad p_{li}^{(n)}(d\theta_2 \times ds_2) = 1;$

(4) $p_{kj}^{(n)}(d\theta_1 \times ds_1) = 1, \quad p_{li}^{(n)}(d\theta_2 \times ds_2) = 1,$

$d\theta_1$, $d\theta_2$ are sets on $\mathfrak{C}$ containing the points θ_1 and θ_2, respectively, for which $m(d\theta_1)$ and $m(d\theta_2)$ are sufficiently small, and ds_1 and ds_2 are small intervals containing s_1 and s_2, respectively. Then, with probability

$$1 - o\left(\frac{m(d\theta_1)\, m(d\theta_2)}{n^2}\, ds_1 ds_2\right)$$

the following equation is satisfied:

$$\begin{aligned}
\xi = {} & \xi(k, j, l, i, 0, 0) + \\
& + [\xi(k, j, l, i, 1, 0) - \xi(k, j, l, i, 0, 0)]\, p_{kj}^{(n)}(d\theta_1 \times ds_1) + \\
& + [\xi(k, j, l, i, 0, 1) - \xi(k, j, l, i, 0, 0)]\, p_{li}^{(n)}(d\theta_2 \times ds_2) + \\
& + [\xi(k, j, l, i, 1, 1) - \xi(k, j, l, i, 0, 0) - \\
& - \xi(k, j, l, i, 1, 0) - \xi(k, j, l, i, 0, 1)]\, p_{kj}^{(n)}(d\theta_1 \times \\
& \times ds_1)\, p_{li}^{(n)}(d\theta_2 \times ds_2).
\end{aligned}$$

Hence

$$\begin{aligned}
\mathsf{M}\xi q_{kj}^{(n)}(d\theta_1 \times ds_1)\, q_{li}^{(n)}(d\theta_2 \times ds_2) = {} & \mathsf{M}[\xi(k, j, l, 1, 1) + \\
& + \xi(k, j, l, i, 0, 0) - \xi(k, j, l, i, 1, 0) - \\
& - \xi(k, j, l, i, 0, 1)]\, \frac{m(d\theta_1)\, m(d\theta_2)}{n^2}\, ds_1 ds_2 +
\end{aligned}$$

$$+ o\left(\frac{m(d\theta_1)\, m(d\theta_2)}{n^2} ds_1 ds_2\right).$$

We denote the limiting value of the random variables

$$[\xi(k, j, l, i, 1, 1) + \xi(k, j, l, i, 0, 0) - \xi(k, j, l, i, 1, 0) - \xi(k, j, l, i, 0, 1)],$$

when $d\theta_1$ tends to θ_1, $d\theta_2$ to θ_2, ds_1 to s_1, ds_2 to s_2, by $\nabla\xi(\theta_1, \theta_2, s_1, s_2)$. In order to calculate this random variables we must substitute in the expression for ξ in terms of $z_r^{(n)}(u)$ the functions $z_r^{(n)}(0, 0, u)$, $z_r^{(n)}(0, 1, u)$, $z_r^{(n)}(1, 0, u)$, $z_r^{(n)}(1, 1, u)$, where $z_r^{(n)}(0, 0, u) = z_r^{(n)}(1, 1, u)$ satisfies equation (1), §3, for $u \neq s_1$, $u \neq s_2$, and moreover,

$$z_k^{(n)}(s_1) - z_k^{(n)}(s_1 - 0) = f(\theta_1, z_k^{(n)}(s_1 - 0), z_j^{(n)}(s_1 - 0)), \tag{16}$$

$$z_k^{(n)}(s_2) - z_k^{(n)}(s_2 - 0) = f(\theta_2, z_k^{(n)}(s_2 - 0), z_j^{(n)}(s_2 - 0)),$$

$$z_l^{(n)}(s_2) - z_l^{(n)}(s_2 - 0) = f(\theta_1, z_l^{(n)}(s_2 - 0), z_i^{(n)}(s_2 - 0)),$$

$$z_i^{(n)}(s_2) - z_i^{(n)}(s_2 - 0) = f(\theta_2, z_i^{(n)}(s_2 - 0), z_l^{(n)}(s_2 - 0)), \tag{17}$$

$z_r^{(n)}(0, 1, u)$ satisfies the same system for $u \neq s_2$ and the relations (17), and $z_r^{(n)}(1, 0, u)$ satisfies the same system for $u \neq s_1$ and the relations (16). The random variables $\nabla\xi(\theta_1, \theta_2, s_1, s_2)$ will be the sum of the first results of the substitution minus the sum of two others: i.e. if $\xi = F(z_1^{(n)}(\cdot), \ldots)$, then

$$\nabla\xi(\theta_1, \theta_2, s_1, s_2) = F(z_1^{(n)}(0, 0, \cdot), \ldots) + F(z_1^{(n)}(1, 1, \cdot), \ldots) - - F(z_1^{(n)}(1, 0, \cdot), \ldots) - F(z_1^{(n)}(0, 1, \cdot), \ldots).$$

Then

$$\mathsf{M}\xi q_{kj}^{(n)}(d\theta_1 \times ds_1)\, q_{li}^{(n)}(d\theta_2 \times ds_2) = = \mathsf{M}\nabla\xi(\theta_1, \theta_2, s_1, s_2)\, \frac{m(d\theta_1)\, m(d\theta_2)\, ds_1 ds_2}{n^2}.$$

Let $s_1 < s_2$. We introduce the random variables

$$\nabla_1\xi(\theta_1, \theta_2, s_1, s_2) = F(z_1^{(n)}(0, 1, \cdot), \ldots) - F(z_1^{(n)}(0, 0, \cdot), \ldots),$$

$$\nabla_2\xi(\theta_1, \theta_2, s_1, s_2) = F(z_1^{(n)}(1, 1, \cdot), \ldots) - F(z_1^{(n)}(1, 0, \cdot) \ldots)$$

(the arguments of the function F in the right members coincide on $[0, s_2)$). Obviously,

$$\mathsf{M}|\nabla\xi(\theta_1, \theta_2, s_1, s_2)| \leqslant \\ \leqslant \mathsf{M}|\nabla_1\xi(\theta_1, \theta_2, s_1, s_2)| + \mathsf{M}|\nabla_2\xi(\theta_1, \theta_2, s_1, s_2)|.$$

Let $\xi = \Psi_{kj}(\varphi, \theta_1 s_1)\Psi_{li}(\varphi, \theta_2, s_2)$. We shall now show how to estimate

$$\mathsf{M}\nabla_m\xi(\theta_1, \theta_2, s_1, s_2), \ m = 1, 2.$$

Both of these quantities are estimated the same way - we shall treat $\nabla_2\xi$. We denote by $U_s^t(n, \omega, 1, 1)$ the operator $U_s^t(n, \omega)$. If in its expression in terms of $z_r^{(n)}(\cdot)$ we substitute $z^{(n)}(1, 1, \cdot)$ in place of these processes, $U_s^t(n, \omega, 1, 0)$ is expressed analogously in terms of $z_r^{(n)}(1, 0, \cdot)$. Obviously,

$$U_{s_1}^{s_2}(n, \omega, 1, 0) = U_{s_1}^{s_2}(n, \omega, 1, 1).$$

Therefore,

$$\begin{aligned}\nabla_2\xi = {} & Q_\theta[U_{s_1}^t(n, \omega, 1, 1) - U_{s_1}^t]\times \\ & \times\varphi(z_k^{(n)}(s_1 - 0), z_j^{(n)}(s_1 - 0))\, Q_\theta[U_{s_2}^t(n, \omega, 1, 1) - \\ & - U_{s_2}^t]\varphi(z_l^{(n)}(1, 0, s_2 - 0), z_i^{(n)}(1, 0, s_2 - 0)) - \\ & - Q_\theta[U_{s_1}^t(n, \omega, 1, 0) - U_{s_1}^t]\varphi(z_k^{(n)}(s_1 - 0), z_j^{(n)}(s_1 - 0))\times \\ & \times Q_\theta[U_{s_2}^t(n, \omega, 1, 0) - U_{s_2}^t]\varphi(z_l^{(n)}(1, 0, s_2 - 0), \\ & z_i^{(n)}(1, 0, s_2 - 0)) = Q_\theta[U_{s_1}^t(n, \omega, 1, 1) - U_{s_1}^t]\times \\ & \times\varphi(z_k^{(n)}(s_1 - 0), z_j^{(n)}(s_1 - 0))\, Q_\theta[U_{s_2}^t(n, \omega, 1, 1) - \\ & - U_{s_2}^t(n, \omega, 1, 0)]\varphi(z_l^{(n)}(1, 0, s_2 - 0), z_i^{(n)}(1, 0, s_2 - 0)) + \\ & + Q_\theta[U_{s_1}^t(n, \omega, 1, 1) - U_{s_1}^t(n, \omega, 1, 0)]\varphi(z_k^{(n)}(s_1 - 0), \\ & z_j^{(n)}(s_1 - 0))\, Q_\theta[U_{s_2}^t(n, \omega, 1, 0) - U_{s_2}^t]\times \\ & \times\varphi(z_l^{(n)}(1, 0, s_2 - 0), z_i^{(n)}(1, 0, s_2 - 0)).\end{aligned}$$

It follows from estimate (15) that there exists a random variable $\eta_1^{(n)}$, not depending on s, for which $\mathsf{M}(\eta_1^{(n)})^2$ is bounded and

$$\| U_s^t(n, \omega, 1, 1) - U_s^t(n, \omega, 1, 0)\| \leqslant \frac{1}{n}\eta_1^{(n)}.$$

Hence,

$$\begin{aligned}|\nabla_2\xi| \leqslant {} & \frac{1}{n}\|Q_\theta\|^2\eta_1^{(n)}(\|U_{s_1}^t(n, \omega, 1, 1)\varphi - U_{s_1}^t\varphi\| + \\ & + \|U_{s_2}^t(n, \omega, 1, 0)\varphi - U_{s_2}^t\varphi\|) = \\ = {} & \frac{1}{n}\|Q_\theta\|^2\eta_1^{(n)}(\|U_{s_1}^t(n, \omega)\varphi - U_{s_1}^t\varphi\| + \\ & + \|U_{s_2}^t(n, \omega)\varphi - U_{s_2}^t\varphi\| + O\left(\frac{1}{n}\right)\right)\end{aligned}$$

(we have again used here an estimate of type (15)). In precisely the same way,

$$|\nabla_1 \xi| \leqslant \frac{1}{n} \|Q_0\|^2 \eta_2^{(n)} (\|U_{s_1}^t(n, \omega)\varphi - U_{s_1}^t\varphi\| + $$
$$+ \|U_{s_2}^t(n, \omega)\varphi - U_{s_2}^t\varphi\| + O\left(\frac{1}{n}\right)).$$

Hence,

$$\frac{1}{n} \sum_{k, j, l, i=1}^{n} \iint\limits_{0<s_1<s_2<t} \iint \mathsf{M}\Psi_{kj}^{(n)}(\varphi, \theta_1, s_1) \times$$
$$\times \Psi_{li}^{(n)}(\varphi, \theta_2, s_2)\, q_{kj}^{(n)}(d\theta_1 \times ds_1)\, q_{li}^{(n)}(d\theta_2 \times ds_2) \leqslant$$
$$\leqslant \frac{1}{n^4} \sum_{k, j, l, i=1}^{n} \iint\limits_{0<s_1<s_2<t} \iint \|Q_0\|^2 \mathsf{M}(\eta_1^{(n)} + \eta_2^{(n)}) \times$$
$$\times (\|U_{s_1}^t(n, \omega)\varphi - U_{s_1}^t\varphi\| + \|U_{s_2}^t(n, \omega)\varphi - U_{s_2}^t\varphi\|) \times$$
$$\times m(d\theta_1)\, m(d\theta_2)\, ds_1 ds_2 + O\left(\frac{1}{n}\right) =$$
$$= O\left(\frac{1}{n} + \left(\int_0^t \mathsf{M}\|U_s^t(n, \omega)\varphi - U_s^t\varphi\|^2 ds\right)^{1/2}\right).$$

Inasmuch as this expression tends to zero as $n \to \infty$, we have the

$$\mathsf{M}[\chi_\varphi^{(n)}(t)]^2 \to 0.$$

This completes the proof of the theorem.

Now suppose we have a system of the form (1), §2. We assume that $\varphi \in C_Z^1$. Then, on the basis of Itô's formula,

$$d\varphi(z_i^{(n)}(t)) = (\varphi'(z_i^{(n)}(t)),\ A(z_i^{(n)}(t))) +$$
$$+ \frac{1}{n} \sum_{j=1}^{n} a(z_i^{(n)}(t),\ z_j^{(n)}(t))\, dt + \frac{1}{n} \sum_{j=1}^{n} \int_\Theta [\varphi(z_i^{(n)}(t) +$$
$$+ f(\theta,\ z_i^{(n)}(t),\ z_j^{(n)}(t))) - \varphi(z_i^{(n)}(t))]\, m(d\theta)\, dt +$$
$$+ \sum_{j=1}^{n} \int_\Theta [\varphi(z_i^{(n)}(t) + f(\theta,\ z_i^{(n)}(t),\ z_j^{(n)}(t))) -$$
$$- \varphi(z_i^{(n)}(t))]\, q_{ij}^{(n)}(d\theta \times dt).$$

Consequently,

$$d\int \varphi(z)\, \mu_t^{(n)}(dz) = \left\{\int (\varphi'(z),\ A(z) + \right.$$
$$+ \int a(z,\ z')\, \mu_t^{(n)}(dz'))\, \mu_t^{(n)}(dz) + \iiint [\varphi(z + f(\theta,\ z,\ z')) -$$
$$- \varphi(z)]\, m(d\theta)\, \mu_t^{(n)}(dz)\, \mu_t^{(n)}(dz')\}\, dt + d\eta_n(\varphi,\ t),$$

where the last term on the right is defined by equation (3). We shall assume that the conditions of Theorem 1, §4, are fulfilled, and hence the limiting statistical distribution function $\lambda_t(dz)$ satisfying equation (3), §4, exists. Subtracting this equation from equation (18), and grouping terms and multiplying by $\sqrt{n}$, we obtain

$$d\int\varphi(z)\nu_t^{(n)}(dz)=\int(\varphi'(z),\ a_n(t,\ z))\nu_t^{(n)}(dz)+ \\ +\int(\varphi'(z'),\ a(z,\ z'))\lambda_t(dz')\nu_t^{(n)}(dz)+ \\ +\int S_t^{(n)}(\omega)\varphi(z)\nu_t^{(n)}(dz)+d\eta_n(\varphi,\ t),$$

where the bounded random operation $S_t^{(n)}(\omega)\varphi$ on C_Z is defined by equation (10), and

$$a_n(t,\ z)=A(z)+\int a(z,\ z')\mu_t^{(n)}(dz).$$

Suppose we have a family of random operators $V_s^t(n,\ \omega)$ on C_Z for which, with $\varphi\in C_Z^1$ and $V_s^t(n,\ \omega)\varphi\in C_Z^1$ and for $\varphi\in C_Z^1$, we have

$$\frac{d}{ds}V_s^t(n,\ \omega)\varphi(z)=-([V_s^t(n,\ \omega)\varphi]'(z),\ a_n(z,\ t))- \\ -R_s^{(n)}(\omega)V_s^t(n,\ \omega)\varphi(z),\ s<t;\ V_t^t(n,\ \omega)\varphi(z)=\varphi(z), \tag{19}$$

where

$$R_s^{(n)}(\omega)\varphi(z)=\int(\varphi'(z'),\ a(z,\ z'))\lambda(dz')+S_s^{(n)}(\omega)\varphi(z). \tag{20}$$

Moreover, suppose we have a family of bounded operators V_s^t on C_Z for which, with $\varphi\in C_Z^1$, we have

$$\frac{d}{ds}V_s^t\varphi(z)=-([V_s^t\varphi]'(z),\ \bar{a}(t,\ z))-\int([V_s^t\varphi]'(z), \\ a(z,\ z'))\lambda_t(dz')-S_sV_s^t\varphi(z),\ s<t,\ V_t^t\varphi(z)=\varphi(z). \tag{21}$$

If, for all $\varphi\in C_Z^1$, $||V_s^t(n,\ \omega)\varphi(z)-V_s^t\varphi(z)||\to 0$, and $||V_s^t(n,\ \omega)||$ are uniformly bounded, then replacing $U_s^t(n,\ \omega)$ in the proof of Theorem 1 by $V_s^t(n,\ \omega)\varphi$, $U_s^t\varphi$ by $V_s^t\varphi$, and the operator $S_z^{(n)}(\omega)$ by the operator

$$S_s^{(n)}(\omega)\varphi(z)+\int(\varphi'(z'),\ a(z,\ z'))\lambda_t(dz')+(\varphi'(z),\ a_n(t,\ z)),$$

we obtain that for all $\varphi\in C_Z^1$ the distribution of the quantity $\int\varphi(z)\nu_t^{(n)}(dz)$ converges to teh ditribution of the quantity

$$\nu_t(\varphi)=\nu_0(V_0^t\varphi)+\int_0^t\iiint Q_\theta V_s^t\varphi(z,\ z')\,\gamma(d\theta\times dz\times dz'\times ds), \tag{22}$$

where the Guassian measure with independent values is the same as in equation (9).

We introduce one sufficient condition under which the assertion just formulated is fulfilled.

LEMMA. Suppose the conditions of Theorem 1, §4, and of Theorem 1, §5, are satisfied and that a(z, z') and A(z) are bounded and continuously differentiable with respect to z and that the operator

$$B_t\varphi(z)=\int(\varphi'(z'),\ a(z,\ z'))\,\lambda_t(dz')$$

can be extended as a bounded operator from C_Z^1 onto $\hat{C}_Z$. Then there exist random operators $V_s^t(n,\ \omega)$ satisfying equation (19), $||V_s^t(n,\ \omega)||$ is uniformly bounded with respect to n and 0 < s < t for every fixed t, and for all $\varphi \in \hat{C}_Z$ we have

$$\|V_s^t(n,\ \omega)\varphi - V_s^t\varphi\|\to 0.$$

Proof. We shall seek operators $V_s^t(n,\ \omega)$ of the form $V_s^t(m,\ \omega) = W_s^t(n,\ \omega)Y_s^t(n,\ \omega)$, where

$$\frac{d}{ds}W_s^t(n,\ \omega)\varphi=-([W_s^t(n,\ \omega)\varphi]'(z),\ a_n(t,\ z)), \qquad s<t,\ W_t^t(n,\ \omega)\varphi=\varphi. \tag{23}$$

Equation (23) can be solved in the following way. We denote by $\zeta_s^t(n,\ z)$ solution of the ordinary differentiable equation in Z:

$$\frac{d}{ds}\zeta_s^t(n,\ z)=a_n(s,\ \zeta_s^t(n,\ z)),\quad s<t,\quad \zeta_t^t(n,\ z)=z. \tag{24}$$

Its solution exists, is unique, is continuously differentiable with respect to z inasmuch as a_n is bounded and continuously differentiable with respect to z. If $W_s^t(n,\ \omega)\varphi = u_n(s,\ t,\ z,\ \omega)$, then

$$\frac{\partial}{\partial s}u_n(s,\ t,\ z,\ \omega)+\left(\frac{\partial}{\partial z}u_n(s,\ t,\ z,\ \omega),\ a_n(s,\ z)\right)=0.$$

Consequently,

$$\frac{d}{ds}u_n(s, t, \zeta_s^t(n, \omega), \omega) =$$
$$= \frac{\partial}{\partial s}u_n(s, t, z, \omega) + \left(u_n'(s, t, z, \omega), \frac{d}{ds}\zeta_s^t(n, z)\right) = 0$$

by virtue of equation (24). Hence,

$$u_n(s, t, \zeta_s^t(n, \omega), \omega) = u_n(t, t, \zeta_t^t(n, z)) = \varphi(z).$$

whence

$$u(s, t, z, \omega) = \varphi([\zeta_s^t]^{-1}(n, z)),$$

where $[\zeta_s^t]^{-1}(n, z)$ is the function inverse to $\zeta_s^t(n, z)$. It is easy to convince oneself that it coincides with the solution of the equation

$$\frac{d}{dz}\eta_s^t(n, z) = -a_n(t, \eta_s^t(n, z)), \quad t > s, \quad \eta_s^s(n, z) = z.$$

Thus, $W_s^t(n, \omega)\varphi(z) = \varphi(\eta_s^t(n, z))$. From this it follows that $||W_s^t(n, \omega)|| = 1$. The operator $W_s^t(n, \omega)$ is invertible: since $\eta_s^t(n, \zeta_s^t(n, z)) = z$, then $[W_s^t(n, \omega)]^{-1}\varphi(z) = \varphi(\zeta_s^t(n, z))$. It follows from this formula that also $||(W_s^t(n, \omega))^{-1}|| = 1$. Finally, it is easy to convince oneself that for each c,

$$\lim_{n\to\infty} \sup_{|z|\leqslant c} [\,|\zeta_s^t(n, z) - \zeta_s^t(z)| + |\eta_s^t(n, z) - \eta_s^t(z)|\,] = 0,$$

where

$$\frac{d}{ds}\zeta_s^t(z) = \bar{a}(s, \zeta_s^t(z)), \qquad s < t, \quad \zeta_t^t(z) = z,$$
$$\frac{d}{d\eta}\eta_s^t(z) = -\bar{a}(t, \eta_s^t(z)), \quad t > s, \quad \eta_s^s(z) = z.$$

This follows from the fact that $a_n(t, z)$ tends uniformly to $\bar{a}(t, z)$ on each bounded set. We set V = WY in equation (19). We then obtain a relation of the form

$$W_s^t(n, \omega)\frac{d}{ds}Y_s^t(n, \omega)\varphi = (B_s + S_s^{(n)}(\omega))\,W_s^t(n, \omega)\,Y_s^t(n, \omega)\,\varphi,$$

whence

$$\frac{d}{ds}Y_s^t(n, \omega)\varphi = [W_s^t(n, \omega)]^{-1}(B_s + S_s^{(n)}(\omega))\,W_s^t(n, \omega)\,Y_s^t(n, \omega)\varphi.$$

Now the proof of the lemma follows from the fact that the operators,

$$[W_s^t(n,\ \omega)]^{-1}(B_s + S_s^{(n)}(\omega))W_s^t(n,\ \omega)$$

are bounded, transform $\hat{C}_Z$ into $\hat{C}_Z$, and for each $\varphi \in \hat{C}_Z$ we have

$$\|[W_s^t(n,\ \omega)]^{-1}(B_s + S_s^{(n)}(\omega))W_s^t(n,\ \omega)\varphi - \\ -(W_s^t)^{-1}(B_s + S_s)W_s^t\varphi\| \to 0$$

as $n \to \infty$.

The next theorem follows from the lemma just proved and our preceding discussion.

THEOREM 2. Suppose the following four conditions are fulfilled: (1) the functions A(z) adn a(z, z') are bounded, satisfy the condition

$$|A(z) - A(\bar{z})| + |a(z,\ z') - a(\bar{z},\ \bar{z}')| \leqslant \\ \leqslant K(|z - \bar{z}| + |\bar{z}' - z'|)$$

for some K, and there exist continuous derivatives $A_z'(z)$, $a_z'(z,\ z')$; (2) $\int|f(\theta,\ z,\ z')|m(d\theta)$ is bounded, for some K we have

$$\int|f(\theta,\ z,\ z') - f(\theta,\ \bar{z},\ \bar{z}')|m(d\theta) \leqslant K(|z - \bar{z}| + |z' - \bar{z}'|)$$

and, for all $\varphi \in \hat{C}_Z$

$$\int\varphi(z + f(\theta,\ z,\ z'))m(d\theta)\mu(dz') \in \hat{C}_Z,$$

where μ is an arbitrary measure on Z; (3) the operator

$$B_t(\varphi) = \int(\varphi'(z'),\ a(z,\ z'))\lambda_t(dz'),\ \text{где}\ \lambda_t(dz)$$

where $\lambda_t(dz)$ is the limiting statistical distribution function, can be extended from C_Z^1 onto $\hat{C}_Z$ and transforms $\hat{C}_Z$ into $\hat{C}_Z$ as a bounded operator; (4) for all $\varphi \in C_Z$,

$$\lim_{n\to\infty}\int\varphi(z)\nu_0^{(n)}(dz).$$

exists.

Then for each $\varphi \in \hat{C}_Z$ the random variable $\int\varphi(z)\nu_t^{(n)}(dz)$ has a limit distribution that coincides with the distribution of the random variable (22).

6. Limit Theorem for Functionals

We shall now consider a system of the form (1), §2, under the conditions of Theorem 1, §4. In §4 we studied the limiting behavior of the statistical distribution function $\mu_t^{(n)}$. This function gives only a representation of the distribution of particles in phase space at each moment of time and does not enable one to judge the nature of the motion of the particles. Theorems 2 and 3, §4, describe the limiting motion of equal individual particle, but also does not allow one to judge the motion of the collection as a whole. In this section we consider results relating to the study of the limiting behavior of the system.

Suppose $D_{[0,T]}(Z)$ is the space of functions without discontinuities of the second kind (defined in §4) and that $G(z(\cdot))$ is a function which is bounded in the metric ρ_D. We shall be interested in the limiting behavior of the mean

$$\frac{1}{n}\sum_{i=1}^{n} G(z_i^{(n)}(\cdot)), \tag{1}$$

where $z_i^{(n)}$ satisfies the system of equations (1), §2.

THEOREM 1. Suppose the conditions of Theorem 2, §4, are satisfied. Then, as $n \to \infty$, the quantity (1) converges in probability to $\mathsf{M}G(\hat{z}(\cdot))$, where $\hat{z}(t)$ is the solution of the stochastic differential equation

$$d\hat{z}(t) = \bar{a}(t, \hat{z}(t))\,dt + \int_\Theta \int_Z f(\theta, \hat{z}(t), z')\,\bar{p}(d\theta \times dz' \times dt) \tag{2}$$

with initial condition $\hat{z}(0)$ having distribution $\lambda_0(dz)$ and not depending on the Poisson measure $\bar{p}$ defined in Theorem 2, §4.

Proof. We shall first prove that

$$\lim_{n\to\infty} \mathsf{D}\left[\frac{1}{n}\sum_{i=1}^{n} G(z_i^{(n)}(\cdot))\right] = 0. \tag{3}$$

It follows from the corollary to Theorem 4, §4, that, for each $c > 0$,

$$\lim_{n\to\infty} \chi_{\{|z_i^{(n)}(0)|\le c,\ |z_j^{(n)}(0)|\le c\}} [\mathsf{M}G(z_i^{(n)}(\cdot)) \times \\ \times G(z_j^{(n)}(\cdot)) - \mathsf{M}G(z_i^{(n)}(\cdot))\,\mathsf{M}G(z_j^{(n)}(\cdot))] = 0.$$

Therefore, for some c_1,

$$\overline{\lim_{n\to\infty}} \mathsf{D}\left[\frac{1}{n}\sum_{i=1}^{n} G(z_i^{(n)}(\cdot))\right] =$$
$$= \overline{\lim_{n\to\infty}} \frac{1}{n^2}\sum_{i\neq j} [\mathsf{M}G(z_i^{(n)}(\cdot))G(z_j^{(n)}(\cdot)) -$$
$$- \mathsf{M}G(z_i^{(n)}(\cdot))\,\mathsf{M}G(z_i^{(n)}(\cdot))] \leqslant \overline{\lim_{n\to\infty}} \frac{1}{n^2}\sum_{i\neq j} c_1\left(\chi_{\{|z_i^{(n)}(0)|>c\}} +\right.$$
$$\left. + \chi_{\{|z_j^{(n)}(0)|>c\}}\right) = 2c_1 \overline{\lim_{n\to\infty}} \mu_0^{(n)}(\{z\colon |z| > c\}).$$

Since, by an appropriate choice of c, the right member can be made arbitrarily small, (3) holds. By the same token, it has been established that the quantity (1) has a limit if, and only if, the limit

$$\lim_{n\to\infty} \frac{1}{n}\sum_{i=1}^{n} \mathsf{M}G(z_i^{(n)}(\cdot)) \tag{4}$$

exists and the limit of the quantity (1) coincides with (4). Suppose the measure ν_n on $D_{[0,T]}(Z)$ is defined by the equation

$$\int G(z)\,\nu_n(dz) = \frac{1}{n}\sum_{i=1}^{n} \mathsf{M}G(z_i^{(n)}(\cdot))$$

for every bounded function G that is continuous in the metric ρ_D. Then

$$\nu_n = \frac{1}{n}\sum_{i=1}^{n} \mu_i^{(n)},$$

where the measure $\mu_i^{(n)}$ on $D_{[0,T]}(Z)$ corresponds to the process $z_i^{(n)}(t)$. We shall denote by μ_z the measure corresponding to the solution of equation (2) with initial condition $\hat{z}(0) = z$. We denote this solution itself by $\hat{z}(t, z)$. Inasmuch as for some c we have

$$|\bar{a}(t, z) - \bar{a}(t, z_1)| +$$
$$+ \int |f(\theta, z, z') - f(\theta, z_1, z')|\, m(d\theta) \leqslant c|z - z_1|,$$

then

$$\sup_{s\leqslant t} |\hat{z}(s, z) - \hat{z}(t, z_1)| \leqslant$$
$$\leqslant |z - z_1| + \int_0^t |\bar{a}(s, \hat{z}(s, z)) - \bar{a}(s, \hat{z}(s, z_1))|\, ds +$$

$$+\int_0^t \iint |f(\theta, \hat{z}(s, z), z') - f(\theta, \hat{z}(s, z_1), z')| p(d\theta \times dz' \times ds),$$

$$\mathsf{M} \sup_{s \leqslant t} |\hat{z}(s, z) - \hat{z}(s, z_1)| \leqslant$$

$$\leqslant |z - z_1| + c\mathsf{M} \int_0^t |\hat{z}(s, z) - \hat{z}(s, z_1)| ds.$$

Consequently,

$$\mathsf{M} \sup_{s \leqslant t} |\hat{z}(s, z) - \hat{z}(s, z_1)| \leqslant |z - z_1| e^{cT}.$$

From this it follows that

$$\int G(z(\cdot)) \mu_s(dz(\cdot)) = \mathsf{M} G(\hat{z}(\cdot, z))$$

is continuous with respect to z. It follows from Theorem 3, §4, that

$$\lim_{n \to \infty} |\mathsf{M} G(z_i^{(n)}(\cdot)) - \mathsf{M} G(\hat{z}(\cdot, z_i^{(n)}(0))| = 0,$$

if $|z_i^{(n)}(0)| \leqq c_1$, where c_1 is an arbitrary constant. We have that

$$\int G(z(\cdot)) \nu_n(dz(\cdot)) = \frac{1}{n} \sum_{i=1}^{n} \int G(z(\cdot)) \mu_i^{(n)}(dz(\cdot)) =$$

$$= \frac{1}{n} \sum_{i=1}^{n} \mathsf{M} G(z_i^{(n)}(\cdot)) = \frac{1}{n} \sum_{i=1}^{n} \mathsf{M} G(\hat{z}(\cdot, z_i^{(n)}(0))) +$$

$$+ \frac{1}{n} \sum_{i=1}^{n} [\mathsf{M} G(z_i^{(n)}(\cdot)) - \mathsf{M} G(\hat{z}(\cdot, z_i^{(n)}(0))] =$$

$$= \int \mathsf{M} G(\hat{z}(\cdot, z)) \mu_0^{(n)}(dz) +$$

$$+ \frac{1}{n} \sum_{i=1}^{n} [\mathsf{M} G(z_i^{(n)}(\cdot)) - \mathsf{M} G(\hat{z}(\cdot, z_i^{(n)}(0)))] \chi_{\{|z_i^{(0)}(0)| \leqslant c_1\}} +$$

$$+ O(\mu_0^{(n)}(\{z: |z| > c_1\})).$$

It follows from the continuity of $\mathsf{M}G(\hat{z}(\cdot, z))$ with respect to z that the first term on the right tends to $\int \mathsf{M}G(\hat{z}(\cdot, z))\lambda_0(dz)$, the second term, as was shown above, tends to zero, and with an appropriate choice of c_1 the third term can be made arbitrarily small for all n. Thus,

$$\lim_{n \to \infty} \int G(z(\cdot)) \nu_n(dz(\cdot)) =$$

$$= \int \mathsf{M} G(\hat{z}(\cdot, z)) \lambda_0(dz) = \iint G(z(\cdot)) \mu_s(dz(\cdot)) \lambda_0(dz).$$

The right member is the limit of the quantity (4) and hence the quantity (3) converges in probability to the right member of the preceding equation. It remains to note that

$$MG(\hat{z}(\cdot)) = \iint G(z(\cdot)) \times \mu_z(dz(\cdot))\lambda_0(dz).$$

This completes the proof of the theorem.

When investigating the interaction of particles one must consider functionals depending on the trajectories of several particles. Suppose $G(z_1(\cdot), \ldots, z_k(\cdot))$ is a function on $D_{[0,T]}(Z^k)$ that is continuous in the metric ρ_D. We consider the quantity

$$\frac{1}{n^k} \sum_{i_1, \ldots, i_k=1}^{n} G(z_{i_1}^{(n)}(\cdot), \ldots, z_{i_k}^{(n)}(\cdot)).$$

Making use of Theorem 4, §4, one can prove the following assertion exactly as one proved Theorem 1.

THEOREM 2. Suppose the conditions of Theorem 1, §4, are fulfilled. Then as $n \to \infty$, the quantity (5) converges in probability to the quantity $MG(\hat{z}_1(\cdot), \ldots, \hat{z}_k(\cdot))$, where $\hat{z}_1(t), \ldots, \hat{z}_k(t)$ are mutually independent processes each of which is distributed as the process $\hat{z}(t)$ defined by equation (2).

We introduce some examples of application of the theorem just formulated. Suppose that in the phase space Z one has selected a region S and that one is interested in the number of particles that are in this region. We denote the boundary of the region S by S'. We assume that the probability of the event at some moment $\tau \leq T$, '$z(\tau) \in S'$ but $\hat{z}(t) \in S \cup S'$ for all $t \leq T$ or $\hat{z}(t) \in X \setminus S$ for all $t \leq T\}$ equals zero. Then the function $G(z(\cdot)) = \{^1_0$ if $z(t) \in S$ for some $t \leq T$; otherwise is continuous with respect to $z(\cdot)$ in ρ_D at all points $z(\cdot) \in D_{[0,T]}(Z)$ except at those for which $G(z(\cdot)) = 0$, but $z(t)$ occurs in S' at some point. Under these assumptions $G(z(\cdot))$ is continuous almost everywhere with respect to the measure $\hat{\mu}$ corresponding to the process $\hat{z}(t)$ on $D_{[0,T]}(Z)$. As is well known (see [15, Vol. 1, p.437, Lemma]), if the sequence of measures μ_n converges weakly to the measure $\hat{\mu}$ in $D_{[0,T]}(Z)$, then the distribution $G(z(\cdot))$ converges in the measure $\hat{\mu}_n$ to the distribution $G(z(\cdot))$ in the measure $\hat{\mu}$ for all G which are continuous (in the metric ρ_D) almsot everywhere in the measure $\hat{\mu}$. Hence, if ψ_n is the number of particles occurring in the region S during the time [0, T], then under the assumptions just made

$$\frac{\psi_n}{n} = \frac{1}{n}\sum_{i=1}^{n} G(z_i^{(n)}(\cdot))$$

converges to the quantity $MG(\hat{z}(\cdot))$ which coincides with the probability of the event that the process $\hat{z}(t)$ occurs in the region S at some moment.

We now assume that a collision occurred if $z_i(t) - z_j(t) \in V$ at some moment of time t, where V is some region in Z. We define the functions $G(z_1(\cdot), z_2(\cdot))$ on $D_{[0,T]}(Z^2)$:

$$G(z_1(\cdot), z_2(\cdot)) = \begin{cases} 1 \text{ if } z_1(t) - z_2 \in V \text{ for some } t \leq T; \\ 0 \text{ otherwise.} \end{cases}$$

The number of 'collisions' equals

$$\sum_{i \neq j} G(z_i^{(n)}(\cdot),\ z_j^{(n)}(\cdot)).$$

If the probability of the event $\{G(\hat{z}_1(\cdot), z_2(\cdot)) = 0$ and, for some $t \leq T$, $\hat{z}_1(t) - \hat{z}_2(t) \in V\}$ equals zero, then the functional G is almost everywhere continuous with respect to the measure μ^2 corresponding to the limit process. Therefore, on the basis of Theorem 2,

$$\frac{1}{n^2}\sum_{i \neq j} G(z_i^{(n)}(\cdot),\ z_j^{(n)}(\cdot)) \to MG(\hat{z}_1(\cdot),\ \hat{z}_2(\cdot))$$

in probability. The quantity on the right corresponds with the probability of the event that the difference of the two independent processes $\hat{z}_1(t) - \hat{z}_2(t)$, each of which is a distribution as $\hat{z}(t)$ occurs at some moment in the region V.

7. Diffusion Approximation

We shall now reformulate the results just obtained for systems of equations of the form (4), §1. With the independence of the coefficients of the equation on n indicated earlier, such a system will take on the form

$$dv_i^{(n)}(t) = \Big[A(x_i^{(n)}(t),\ v_i^{(n)}(t)) + + \frac{1}{n}\sum_{j=1}^{n} a(x_i^{(n)}(t),\ x_j^{(n)}(t),\ v_i^{(n)}(t),\ v_j^{(n)}(t)) \Big] dt + \tag{1}$$

$$+ \sum_{j=1}^{n} \int_{\Theta} f(\theta,\ x_i^{(n)}(t),\ v_i^{(n)}(t),\ x_j^{(n)}(t),\ v_j^{(n)}(t))\, p_{ij}^{(n)}(d\theta \times dt); \quad (1)$$

here $x_i^{(n)}(t)$ is the radius vector of the i-th particle at the moment t and $v_i^{(n)}(t)$ is its velocity.

Suppose the following conditions are fulfilled:

(1) For some $c > 0$, $\alpha > 1$, $0 < \alpha_1 < \alpha$

$$|A(x, v) - A(\bar{x}, \bar{v})| \leqslant c(|x - \bar{x}| + |v - \bar{v}|),$$
$$|A(x, v)| \leqslant c(1 + |x|^{\alpha_1} + |v|),$$
$$|a(x, y, v, u)| \leqslant c(1 + |x| + |y| + |u| + |v|),$$
$$|a(x, y, v, u) - a(\bar{x}, \bar{y}, \bar{v}, \bar{u})| \leqslant$$
$$\leqslant c(|x - \bar{x}| + |y - \bar{y}| + |v - \bar{v}| + |u - \bar{u}|),$$
$$\int_\Theta |f(\theta, x, v, y, u) - f(\theta, \bar{x}, \bar{v}, \bar{y}, \bar{u})|^\alpha m(d\theta) \leqslant$$
$$\leqslant c(|x - \bar{x}|^\alpha + |y - \bar{y}|^\alpha + |v - \bar{v}|^\alpha + |u - \bar{u}|^\alpha);$$
$$\int_\Theta |f(\theta, x, v, y, u)|^\alpha m(d\theta) \leqslant c(1 + |x|^\alpha + |v|^\alpha + |y|^\alpha + |u|^\alpha).$$

(2) We denote by $\mu_i^{(n)}(dx, dv)$ the statistical distribution function of the system: for each bounded continuous function $\varphi(x, v)$,

$$\int \varphi(x, v)\mu_t^{(n)}(dx, dv) = \frac{1}{n}\sum_{i=1}^{n} \varphi(x_i^{(n)}(t), v_i^{(n)}(t)),$$

the sequence of measures $\mu_t^{(n)}(dx, dv)$ converges weakly to the measure $\lambda_0(dx, dv)$ and

$$\overline{\lim_{n\to\infty}} \int (|x|^{\alpha_1} + |v|^{\alpha_1})\mu_0^{(n)}(dx, dv) < \infty.$$

If these conditions are fulfilled, then the conditions of Theorem 1, §4, are fulfilled with the functions

$$\rho(z) = \rho(x, v) = |x|^{\alpha_1} + |v|^{\alpha_1}, \quad \lambda(|z|) = |z|^\alpha.$$

Therefore, the following assertions that follow from Theorems 1 and 3 of §4 will be valid.

I. For all $t > 0$, the statistical distribution function $\mu_t^{(n)}(dx, dv)$ converges weakly in probability to the nonrandom probability measure $\lambda_t(dx, dv)$, whereby, for all $\varphi \in C_{x^2}$, $\int\varphi(x, v)\lambda_t(dx, dv)$ is continuous with respect to $t > 0$, $t \geq 0$ and for $t > 0$ for every function $\varphi(x, v) \in C_{x^2}$ and continuously differentiable with respect to x, the following relation is fulfilled:

$$\begin{aligned}\frac{d}{dt}\int \varphi(x, v)\lambda_t(dx, dv) = \\ = \int\Big[(\varphi'_x(x, v), v) + (\varphi'_v(x, v), A(x, v)) + \\ + \int a(x, y, v, u)\lambda_t(dy, du)\Big]\lambda_t(dx, dv) + \\ + \iiint [\varphi(x, v + f(\theta, x, v, y, u)) - \varphi(x, v)] \times \\ \times m(d\theta)\lambda_t(dy, du)\lambda_t(dx, dv).\end{aligned} \tag{2}$$

The measure $\lambda_t(dx, dv)$ is the limit as $r \to \infty$ of the measures $\lambda_t^{(r)}(dx, dv)$, which are uniquely determined by the relation

$$\begin{aligned}\frac{d}{dt}\int \varphi(x, v)\lambda_t^{(r)}(dx, dv) = \\ = r\iint\Big[\varphi\Big(x + \frac{1}{r}v, v + \frac{1}{r}(A(x, v) + a(x, y, v, u)\Big) - \\ - \varphi(x, v)\Big]\lambda_t^{(r)}(dy, du)\lambda_t^{(r)}(dx, dv) + \\ + \iiint [\varphi(x, v + f(\theta, x, v, y, u)) - \varphi(x, v)] \times \\ \times m(d\theta)\lambda_t^{(r)}(dy, du)\lambda_t^{(r)}(dx, dv).\end{aligned} \tag{3}$$

II. We denote

$$\bar{a}_t(x, v) = A(x, v) + \int a(x, y, v, u)\lambda_t(dy, du),$$

and let $p(d\theta \times dx \times dv \times dt)$ be Poisson measure on $\Theta \times X^2 \times [0, \infty)$ for which

$$\mathsf{M}p(d\theta \times dx \times dv \times dt) = m(d\theta)\lambda_t(dx, dv)\,dt.$$

If $x_1^{(n)}(0) \to x(0)$, $v_1^{(n)}(0) \to v(0)$, then finite-dimensional distributions and distributions of functions that are continuous on $D_{[0,T]}(X^2)$ in the metric ρ_D for the processes $(x_1^{(n)}(t), v_1^{(n)}(t))$ converge to finite-dimensional distributions and distributions of the same functionals for the process $(\bar{x}_1(t), \bar{v}_1(t))$ which is the solution of the stochastic equation

$$\begin{aligned}d\bar{v}_1(t) = \bar{a}_t(\bar{x}_1(t), \bar{v}_1(t))\,dt + \\ + \iiint f(\theta, \bar{x}_1(t), \bar{v}_1(t), x, v)\,p(d\theta \times dx \times dv \times dt)\end{aligned} \tag{4}$$

with initial condition $\bar{v}_1(0) = v_1(0)$, where

$$\bar{x}_1(t) = x_0 + \int_0^t \bar{v}_1(s)\,ds.$$

In this section we shall investigate equation (4) under the assumption that the velocity increases indefinitely on account of the growth of $\bar{a}_t$ and the increase in the number of jumps of the jumplike component (growth of the measure m). It turns out in this case, that there is a definite 'braking' - i.e. the mean of the increments of the velocity directed against the velocity - then although v(t) becomes infinite, x(t) remains bounded and in the limit the finite-dimensional distributions of the process x(t) coincide with the finite-dimensional distribtuions of some diffusion process. To some degree this clarifies the situation that equations obtained in probability theory for Brownian motion are first-order equations and the equations of motion of particles (Newton equations) are of the second order.

We shall consider a family of equations of the form (4) with coefficients depending on a small parameter ε and investigate the behavior of the solution as $\varepsilon \to 0$. Suppose $x_\varepsilon(t)$, $v_\varepsilon(t)$ are solutions of the equation

$$dv_\varepsilon(t) = \frac{1}{\varepsilon}(-M(t, x_\varepsilon)v_\varepsilon + a(t, x_\varepsilon) + \alpha(t, x_\varepsilon, v_\varepsilon))\,dt + \\ + \int f_\varepsilon(\theta, x_\varepsilon, v_\varepsilon)\, q_\varepsilon(d\theta \times dt),$$

where $M(t, x)$ is some strictly positive operator in X, $a(t, x) \in X$, $\alpha_\varepsilon(t, x, v) = o(1)$ uniformly with respect to t, x and v, $q_\varepsilon(d\theta \times dt) = p_\varepsilon(d\theta \times dt) - m_\varepsilon(t, d\theta)dt$, where $m_\varepsilon(t, d\theta)dt = Mp_\varepsilon(d\theta \times dt)$, $p_\varepsilon(d\theta \times dt)$ is a Poisson measure on $\Theta \times [0, \infty)$ (clearly, $\Theta \times X^2$ can be taken as a new Θ). Of course, the strict positiveness of the operator $M(t, x)$ means the presence of the required braking. We note that the braking in a viscous medium will be proportional to the velocity and directed opposite to the velocity, i.e. in the given case $M(t, x) = \mu(t, x)E$, where E is the identity operator and μ is a coefficient depending on the properties of the medium at the moment t in the point x.

THEOREM 1. Suppose the following conditions are fulfilled:

(1) $\overline{\lim\limits_{\varepsilon\to 0}} \sup\limits_{x, v, t} \varepsilon^3 \int |f_\varepsilon(\theta, x, v)|^4 m_\varepsilon(t, d\theta) < \infty$;

(2) there exists a $g_\varepsilon(\theta, x)$ such that

$$\lim_{\varepsilon\to 0} \sup_{\theta, x} |g_\varepsilon(\theta, x)| = 0 \; u$$

and

$$\lim_{\varepsilon\to 0} \sup_{x, v, t} \int |\varepsilon f_\varepsilon(\theta, x, v) - g_\varepsilon(\theta, x)|^2 m_\varepsilon(t, d\theta) = 0,$$

$$\lim_{\varepsilon\to 0} \sup_{t, x} \int |g_\varepsilon(\theta, x)|^4 m_\varepsilon(t, d\theta) = 0.$$

(3) there exists a nonnegative operator B(t, x) such that

$$\lim_{\varepsilon\to 0}\sup_{x,v,t}\left|\int (g_\varepsilon(\theta, x), z)^2 m_\varepsilon(t, d\theta) - (B(t, x)z, z)\right| = 0,$$
$$z \in X;$$

(4) the functions a(t, x), B(t, x), M(t, x) are continuous simultaneously in all variables and bounded;

(5) for some $\gamma > 0$ $(M(t, x)v, v) \geq \delta(v, v)$ and the continuous derivative of the operator M(t, x) with respect to t and x exists,

(6) the functions a(t, x), B(t, x) and $\{M(t, x)]'_x$ satisfy a Lipschitz condition with respect to x with a constant not depending on t.

Then: if

$$x_\varepsilon(0) \to x(0)\ u\ |v_\varepsilon(0)| = O\left(\frac{1}{\sqrt{\varepsilon}}\right),$$

then the distribution of the process $x_\varepsilon(t)$ converges weakly in $D_{[0,T]}(X)$ to the distribution of the diffusion process x(t) which is the solution of the stochastic differential equation

$$dx(t) = \tilde{a}(t, x(t))\,dt + \tilde{B}(t, x(t))\,dw(t) \tag{6}$$

with the initial condition x(0); here, w(t) is a Wiener process in X,

$$\left.\begin{aligned} &\tilde{B}(t, x) = (M^{-1}(t, x)B(t, x)M^{*-1}(t, x))^{1/2},\\ &\tilde{a}(t, x) = M^{-1}(t, x)a(t, x) +\\ &\quad + \sum_{j,k,l=1}^{r}(D(t, x)e_k, e_l)([M^{-1}(t, x)]'_x(e_k)e_l, e_j)e_j, \end{aligned}\right\} \tag{7}$$

where

$$D(t, x) = \int_0^\infty e^{-sM(t,x)}B(t, x)e^{-sM^*(t,x)}ds,$$

$\{e_1, \ldots, e_{\bar{yr}}\}$ is an orthonormalized basis in X,

$[M^{-1}]'_x(u)$ is the derivative of the operator M^{-1} in the direction u,

M* is the operator adjoint to M.

<u>Proof</u>. We carry out the proof in several steps.

1. We shall show that the quantity $M\varepsilon(v_\varepsilon(t), v_\varepsilon(t))$ is bounded. In fact, by Itô 's formula,

$$\mathsf{M}d(v_\varepsilon(t), v_\varepsilon(t)) = 2\mathsf{M}\Big\{\frac{1}{\varepsilon}[-M(t, x_\varepsilon(t))v_\varepsilon(t) + a(t, x_\varepsilon(t)) + \alpha_\varepsilon(t, x_\varepsilon(t), v_\varepsilon(t)), v_\varepsilon(t))]\Big\}dt + \int_\Theta |f_\varepsilon(\theta, x_\varepsilon(t), v_\varepsilon(t))|^2 m_\varepsilon(t, d\theta)\,dt.$$

Therefore,

$$\frac{d}{dt}\mathsf{M}|v_\varepsilon(t)|^2 \leqslant -\frac{2\gamma}{\varepsilon}\mathsf{M}|v_\varepsilon(t)|^2 + \frac{c_1}{\varepsilon^2}, \tag{8}$$

where c_1 is some constant (we made use of the boundedness of $a(t, x)$, $\alpha_\varepsilon(t, x, v)$, $B(t, x)$ and of conditions (2), (3), (5)). It follows from inequality (8) that

$$\varepsilon\mathsf{M}|v_\varepsilon(t)|^2 \leqslant \max\Big[\varepsilon|v_\varepsilon(0)|^2, \frac{c_1}{2\gamma}\Big]. \tag{9}$$

2. We shall now show that the quantity $\varepsilon^2\mathsf{M}|v_\varepsilon(t)|^4$ is bounded. Again using Itô's formula, we find that

$$\begin{aligned} d\mathsf{M}|v_\varepsilon(t)|^4 \leqslant\ & 4\mathsf{M}\Big(\frac{1}{\varepsilon}[-M(t, x_\varepsilon(t))v_\varepsilon(t) + a(t, x_\varepsilon(t)) + \\ & + \alpha_\varepsilon(t, x_\varepsilon(t), v_\varepsilon(t))], v_\varepsilon(t)\Big)|v_\varepsilon(t)|^2 dt + \\ & + \mathsf{M}\int_\Theta \{6|f_\varepsilon(\theta, x_\varepsilon(t), v_\varepsilon(t))|^2 \cdot |v_\varepsilon(t)|^2 + \\ & + 4|f_\varepsilon(\theta, x_\varepsilon(t), v_\varepsilon(t))|^3 \cdot |v_\varepsilon(t)| + \\ & + |f_\varepsilon(\theta, x_\varepsilon(t), v_\varepsilon(t))|^4\} m_\varepsilon(t, d\theta)\,dt. \end{aligned} \tag{10}$$

Furthermore, we have

$$\mathsf{M}|v_\varepsilon(t)|^2 \int |f_\varepsilon(\theta, x_\varepsilon(t), v_\varepsilon(t))|^2 m_\varepsilon(t, d\theta) \leqslant \frac{c_1}{\varepsilon^3}\mathsf{M}|v_\varepsilon(t)|^2,$$

$$\int |f_\varepsilon(\theta, x_\varepsilon(t), v_\varepsilon(t))|^4 m_\varepsilon(t, d\theta) \leqslant \frac{c_1}{\varepsilon^3},$$

$$\begin{aligned} \mathsf{M}|v_\varepsilon(t)| & \int |f_\varepsilon(\theta, x_\varepsilon(t), v_\varepsilon(t))|^3 m_\varepsilon(t, d\theta) \leqslant \\ & \leqslant \frac{1}{2}\mathsf{M}|v_\varepsilon(t)|^2 \int |f_\varepsilon(\theta, x_\varepsilon(t), v_\varepsilon(t))|^2 m_\varepsilon(t, d\theta) + \\ & + \frac{1}{2}\mathsf{M}\int |f_\varepsilon(\theta, x_\varepsilon(t), v_\varepsilon(t))|^4 m_\varepsilon(t, d\theta) \leqslant \\ & \leqslant \frac{c_1}{2}\Big(\frac{1}{\varepsilon^3}\mathsf{M}|v_\varepsilon(t)|^2 + \frac{1}{\varepsilon^3}\Big). \end{aligned}$$

Moreover, for all $\gamma_1 > 0$,

$$(a(t, x)+\alpha_\varepsilon, v) \leqslant \gamma_1 |v|^2 + \frac{1}{4\gamma_1} |a(t, x)+\alpha_\varepsilon|^2.$$

Therefore, it follows from inequality (10) that there exists a constant c_2 such that

$$\frac{d}{dt} \mathsf{M} |v_\varepsilon(t)|^4 \leqslant -\frac{\gamma}{\varepsilon} \mathsf{M} |v_\varepsilon(t)|^4 + \frac{c^2}{\varepsilon^3}.$$

Hence,

$$\varepsilon^2 \mathsf{M} |v_\varepsilon(t)|^4 \leqslant \max\left[\varepsilon^2 |v_\varepsilon(0)|^4, \frac{c_2}{\gamma}\right].$$

3. We rewrite equation (5) in the form

$$M(t, x_\varepsilon(t))\, dx_\varepsilon(t) = a(t, x_\varepsilon(t))\, dt + \int \varepsilon f_\varepsilon(\theta, x_\varepsilon(t), v_\varepsilon(t)) \times \times q_\varepsilon(d\theta \times dt) + \alpha_\varepsilon(t, x_\varepsilon(t), v_\varepsilon(t))\, dt - \varepsilon dv_\varepsilon(t).$$

whence

$$x_\varepsilon(t) - x_\varepsilon(0) = \int_0^t M^{-1}(s, x_\varepsilon(s))\, a(s, x_\varepsilon(s))\, ds + \int_0^t \int_\Theta \varepsilon M^{-1}(s, x_\varepsilon(s)) f_\varepsilon(\theta, x_\varepsilon(s), v_\varepsilon(s))\, q_\varepsilon(d\theta \times ds) - \varepsilon \int_0^t M^{-1}(s, x_\varepsilon(s))\, dv_\varepsilon(s) + \alpha_\varepsilon(t), \tag{11}$$

where $\alpha_\varepsilon(t)$ tends to zero uniformly with respect to t. Making use of condition (2) we can convince ourselves that

$$\int_0^t \int_\Theta \varepsilon M^{-1}(s, x_\varepsilon(s)) f_\varepsilon(\theta, x_\varepsilon(s), v_\varepsilon(s))\, q_\varepsilon(d\theta \times ds) = \int_0^t \int_\Theta M^{-1}(s, x_\varepsilon(s))\, g_\varepsilon(\theta, x_\varepsilon(s))\, q_\varepsilon(d\theta \times ds) + \beta_\varepsilon(s),$$

where $\beta_\varepsilon(s)$ is a square integrable martingale whose characteristic tends to zero. Therefore, $\beta_\varepsilon(t)$ converges in probability to zero uniformly on each finite interval.

We shall make use of the differentiability with respect to all variables simultaneously of $M^{-1}(s, x)$ and integrate the integral with respect to dv_ε in equation (11) by parts. We obtain

$$\int_0^t M^{-1}(s, x_\varepsilon(s))\,dv_\varepsilon(s) = M^{-1}(t, x_\varepsilon(t))\,v_\varepsilon(t) -$$
$$- M^{-1}(0, x_\varepsilon(0), v_\varepsilon(0)) - \int_0^t [M^{-1}(s, x_\varepsilon(s))]'_s v_\varepsilon(s)\,ds -$$
$$- \int_0^t [M^{-1}(s, x_\varepsilon(s))]'_x (v_\varepsilon(s))\,v_\varepsilon(s)\,ds$$

(here $[\ \ ]'_s$ is the partial derivative of the operator with respect to s and $[\ \ ]'_x(v)$ is the partial derivative of the operator with respect to x in the direction v and depending linearly on v). Since the operators $M^{-1}(t, x)$ and $[M^{-1}(t, x)]'_t$ are bounded, then $\varepsilon M^{-1}(t, x_\varepsilon(t))v_\varepsilon(t)$, $\varepsilon M^{-1}(t, x_\varepsilon(0))v_\varepsilon(0)$ and

$$\varepsilon \int_0^t [M^{-1}(\varepsilon, x_\varepsilon(s)]'_s v_\varepsilon(s)\,ds$$

tend to zero as $\varepsilon \to 0$. Thus equation (11) can be rewritten in the form

$$x_\varepsilon(t) - x_\varepsilon(0) = \int_0^t M^{-1}(s, x_\varepsilon(s))\,a(s, x_\varepsilon(s))\,ds +$$
$$+ \int_0^t \int_\Theta M^{-1}(s, x_\varepsilon(s))\,g_\varepsilon(\theta, x_\varepsilon(s))\,q_\varepsilon(d\theta \times ds) + \tag{12}$$
$$+ \varepsilon \int_0^t [M^{-1}(s, x_\varepsilon(s))]'_x (v_\varepsilon(s))\,v_\varepsilon(s)\,ds + \delta_\varepsilon(t),$$

where $\delta_\varepsilon(t)$ is a random process which converges in probability to zero uniformly with resepct to t as $\delta \to 0$.

4. We shall show that $\delta_\varepsilon(t)$ in (12) is such that, for all $T > 0$, the quantity $\sup_{t \leq T} |\delta_\varepsilon(t)|$ tends to zero in probability. To this end, it suffices to show that, for all T,

$$\sup_{t \leqslant T} \varepsilon |v_\varepsilon(t)| \to 0 \tag{13}$$

in probability. If in the formula with which we obtained (8) we do not take the mathematical expectation, then we find that

$$d(v_\varepsilon(t), v_\varepsilon(t)) \leqslant$$
$$\leqslant \left[-\frac{2\gamma}{\varepsilon}(v_\varepsilon(t), v_\varepsilon(t)) + \frac{c_1}{\varepsilon^2}\right]dt + \frac{2}{\varepsilon}\int_\Theta (v, \varepsilon f)\,q_\varepsilon(d\theta \times dt).$$

We consider this inequality for $t \in [k\varepsilon, (k+1)\varepsilon]$. We shall have

$$de^{\frac{\gamma}{\varepsilon}t}|v_\varepsilon(t)|^2 \leqslant e^{\frac{\gamma}{\varepsilon}t}\frac{c_1}{\varepsilon^2}dt + \frac{2}{\varepsilon}\int_\Theta e^{\frac{\gamma}{\varepsilon}t}(v, \varepsilon f)\, q_\varepsilon(d\theta \times dt),$$

$$e^{\frac{\gamma}{\varepsilon}t}|v_\varepsilon(t)|^2 \leqslant |v_\varepsilon(k\varepsilon)|^2 e^{\frac{\gamma}{\varepsilon}\varepsilon k} + \frac{c_1}{\varepsilon^2}\int_{k\varepsilon}^{t} e^{\frac{\gamma}{\varepsilon}s}ds + \int_{k\varepsilon}^{t}\int_\Theta e^{\frac{\gamma}{\varepsilon}s}(v, \varepsilon f)\, q_\varepsilon(d\theta \times ds),$$

$$\sup_{k\varepsilon\leqslant t\leqslant(k+1)\varepsilon}|v_\varepsilon(t)|^2 \leqslant e^{\gamma}\left[|v_\varepsilon(k\varepsilon)|^2 + \frac{c_1}{\varepsilon\gamma}(1-e^{-\gamma})\right] + \sup_{k\varepsilon\leqslant t\leqslant(k+1)\varepsilon}\left|\int_{k\varepsilon}^{t}\int_\Theta e^{\frac{\gamma}{\varepsilon}(s-k\varepsilon)}(v, \varepsilon f)\, q_\varepsilon(d\theta \times ds)\right|.$$

If $\psi_\varepsilon(s) = \exp\left\{\frac{\gamma}{\varepsilon}(s - k\varepsilon)\right\}$ for $k\varepsilon < s \leqq (k+1)\varepsilon$, then

$$\sup_{t\leqslant T}|v_\varepsilon(t)|^2 \leqslant \frac{c_1}{\varepsilon\gamma}(e^{\gamma}-1) + e^{\gamma}\sup_{k\leqslant T/\varepsilon}|v_\varepsilon(k\varepsilon)|^2 + 2\sup_{t\leqslant T}\left|\int_0^t\int_\Theta \psi_\varepsilon(s)(v_\varepsilon(s), f_\varepsilon(\theta, x_\varepsilon(s), v_\varepsilon(s)))\, q_\varepsilon(d\theta \times ds)\right|. \tag{14}$$

We note that

$$\mathsf{P}\{\varepsilon^2\sup_{k\leqslant T/\varepsilon}|v_\varepsilon(k\varepsilon)|^2 > \lambda\} \leqslant \sum_{k\leqslant T/\varepsilon}\mathsf{P}\{\varepsilon^2|v_\varepsilon(k\varepsilon)|^2 > \lambda\} \leqslant \sum_{k\leqslant T/\varepsilon}\frac{\mathsf{M}\varepsilon^4|v_\varepsilon(k\varepsilon)|^4}{\lambda^2} \leqslant \frac{T}{\varepsilon\lambda^2}\varepsilon^4\sup_{t\leqslant T}\mathsf{M}|v_\varepsilon(t)|^4 = O(\varepsilon).$$

Furthermore,

$$\mathsf{M}\sup_{t\leqslant T}\left|\int_0^t\int_\Theta \psi_\varepsilon(s)(v_\varepsilon(s), f_\varepsilon(\theta, x_\varepsilon(s), v_\varepsilon(s)))\, q_\varepsilon(d\theta \times ds)\right|^2 \leqslant 4\mathsf{M}\int_0^T\int_\Theta \psi_\varepsilon^2(s)(v_\varepsilon(s), f_\varepsilon(\theta, x_\varepsilon(s), v_\varepsilon(s)))^2 m_\varepsilon(s, d\theta)\, ds \leqslant 4e^{2\gamma}\frac{1}{\varepsilon^2}c_1\mathsf{M}\int_0^T|v_\varepsilon(s)|^2 ds = O(\varepsilon^{-3}).$$

Therefore,

$$\mathsf{M}\varepsilon^2 \sup_{t\leq T}\left|\int_0^t\int_\Theta \psi_\varepsilon(s)(v_\varepsilon(s),\ f_\varepsilon(\theta,\ x_\varepsilon(s)v_\varepsilon(s)))\,q_\varepsilon(d\theta\times ds)\right| = \\ = O(\varepsilon^{1/2}).$$

Thus, after multiplying inequality (14) by ε^2, each term in the right member converges in probability to zero. This completes the proof of relation (13).

5. We shall now show that the family of measures corresponding in $C_{[0,T]}(X)$ to the processes

$$\zeta_\varepsilon(t)=\varepsilon\int_0^t [M^{-1}(s, x_\varepsilon(s))]'_x(v_\varepsilon(s))v_\varepsilon(s)\,ds,$$

is compact. By virtue of the hypothesis placed on the operator M, we have that

$$|[M^{-1}(s,\ x)]'_x(v)v| \leqslant c_3|v|^2,$$

where c_3 is some constant. Therefore, for $0 \leqq t < t + h \leqq T$,

$$\zeta_\varepsilon(t+h)-\zeta_\varepsilon(t)| \leqslant c_3\int_t^{t+h}\varepsilon|v_\varepsilon(s)|^2\,ds \leqslant \\ \leqslant c_3\sqrt{h}\sqrt{\int_0^T \varepsilon^2|v_\varepsilon(s)|^4\,ds}\,.$$

Inasmuch as it was proved in step 2, above, that $\mathsf{M}\int_0^T\varepsilon^2|v_\varepsilon(s)|^4\,ds$ is uniformly bounded, the required conclusion follows from the preceding inequality.

6. The family of measures corresponding to the processes $x_\varepsilon(t)$ in $D_{[0,T]}(X)$ is also compact. In fact, we denote

$$a_1(t,\ x)=M^{-1}(t,\ x)\,a(t,\ x),$$
$$h_\varepsilon(0,\ t,\ x)=M^{-1}(t,\ x)\,g_\varepsilon(t,\ x),$$
$$\eta_\varepsilon(t)=\zeta_\varepsilon(t)+\delta_\varepsilon(t).$$

Then

$$x_\varepsilon(t)-x_\varepsilon(0)= \\ =\int_0^t a_1(s,\ x_\varepsilon(s))\,ds+\int_0^t h_\varepsilon(0,\ s,\ x_\varepsilon(s))\,q_\varepsilon(d0\times ds)+\eta_\varepsilon(t).$$

The boundedness in probability of $\sup_{t \leq T} |x_\varepsilon(t)|$ follows from the boundedness of $a_1(s, x)$ and $\int h_\varepsilon(0, s, x) m_\varepsilon(s, d\theta)$ and the boundedness in probability of $\sup_{t \leq T} |\eta_\varepsilon(T)|$. The functions $\int_0^t a_1(s, x_\varepsilon(s))ds$ are uniformly continuous and the compactness of the integrals follows from the inequality.

$$\mathsf{M}\left[\left|\int_t^{t+h} h_\varepsilon(\theta, s, x_\varepsilon(s))\, q_\varepsilon(d\theta \times ds)\right|^2 \Big/ \mathcal{F}_t^{(\varepsilon)}\right] \leqslant c_4 h,$$

where $F_t^{(\varepsilon)}$ is the σ-algebra generated by the quantities $x_\varepsilon(s)$, $v_\varepsilon(s)$, $s \leq t$, and c_4 is some constant.

7. We shall now consider the limiting behavior of integrals of the form

$$\varepsilon \int_0^t \varphi(x_\varepsilon(s))(v_\varepsilon(s), z)(v_\varepsilon(s), u)\, ds,$$

where $\varphi(x)$ is a bounded continuous function, and z and u belong to X. We shall show that

$$\lim_{\varepsilon \to 0} \mathsf{M}\left|\varepsilon \int_0^t \varphi(x_\varepsilon(s))(v_\varepsilon(s), z)(v_\varepsilon(s), u)\, ds - \right. \\ \left. - \int_0^t \varphi(x_\varepsilon(s))(D(s, x_\varepsilon(s))z, u)\, ds\right| = 0. \tag{15}$$

($D(t, x)$ is defined in the formulation of the theorem.) Inasmuch as $\varphi(x)$ is bounded and continuous, then, for $0 = t_0 < t_1 < \ldots < t_n = t$,

$$\mathsf{M}\left|\varepsilon \int_0^t \varphi(x_\varepsilon(s))(v_\varepsilon(s), z)(v_\varepsilon(s), u)\, ds - \right. \\ \left. - \varepsilon \sum_{k=1}^n \int_{t_{k-1}}^{t_k} \varphi(x_\varepsilon(t_{k-1}))(v_\varepsilon(s), z)(v_\varepsilon(s), u)\, ds\right| \leqslant \\ \leqslant \left(\sum_{k=1}^n \int_{t_{k-1}}^{t_k} \mathsf{M}|\varphi(x_\varepsilon(s)) - \varphi(x_\varepsilon(t_{k-1}))|^2\, ds\right)^{1/2} \times \\ \times \left(\mathsf{M}\int_0^t \varepsilon^2 |v_\varepsilon(s)|^4\, ds\right)^{1/2} |z| \cdot |u|$$

and the right member of this inequality can be made arbitrarily small by the appropriate choice of a sufficiently small $\max_k (t_k - t_{k-1})$ by virtue of the compactness of the measures corresponding to $x_\varepsilon(t)$ in $D_{[0,T]}(X)$. Therefore, to prove (15) it suffices to show that for sufficiently small $h > 0$ we have

$$\lim_{\varepsilon\to 0} \mathsf{M}\left|\varepsilon \int_t^{t+h} (v_\varepsilon(s),\ z)(v_\varepsilon(s),\ u)\,ds - \int_t^{t+h} (D(s,\ x_\varepsilon(s))\,z,\ u)\,ds\right| = 0.$$

But to prove this relation it suffices to show that

$$\overline{\lim_{\varepsilon\to 0}}\, \mathsf{M}\left|\varepsilon \int_t^{t+h} (v_\varepsilon(s),\ z)(v_\varepsilon(s),\ u)\,ds - \int_t^{t+h} (D(s, x_\varepsilon(s))\,z, u)\,ds\right| = o(h) \tag{16}$$

uniformly relative to t. The relation

$$\lim_{h\to 0} \overline{\lim_{\varepsilon\to 0}} \sup_{t\le s\le t+h} \mathsf{M}\,|x_\varepsilon(s) - x_\varepsilon(t)| = 0. \tag{17}$$

follows from equation (12) and estimate (9). Using the continuity and boundedness of $D(t, x)$ and this relation, we find:

$$\overline{\lim_{\varepsilon\to 0}}\, \mathsf{M}\left|\int_t^{t+h} (D(s,\ x_\varepsilon(s))\,z,\ u)\,ds - h(D(t,\ x_\varepsilon(t))\,z,\ u)\right| = o(h). \tag{18}$$

Let $\hat{v}_\varepsilon(t) = v_\varepsilon(t)$ and suppose that for $t \leqq s \leqq t + h$ the equation

$$d\hat{v}_\varepsilon(s) = -\frac{1}{\varepsilon} M(t,\ x_\varepsilon(t))\,\hat{v}_\varepsilon(s) + \frac{1}{\varepsilon}\int_\Theta g_\varepsilon(\theta,\ x_\varepsilon(t))\,q_\varepsilon(d\theta\times ds).$$

is satisfied. Then

$$\begin{aligned} d(v_\varepsilon(s) - \hat{v}_\varepsilon(s)) = &\\ = -\frac{1}{\varepsilon} &M(t,\ x_\varepsilon(t))(v_\varepsilon(s) - \hat{v}_\varepsilon(s))\,ds + \\ +\frac{1}{\varepsilon}&[a(s,\ x_\varepsilon(s)) + \alpha_\varepsilon(s,\ x_\varepsilon(s),\ v_\varepsilon(s))]\,ds + \\ +\frac{1}{\varepsilon}&[M(t,\ x_\varepsilon(t)) - M(s,\ x_\varepsilon(s))]\,v_\varepsilon(s)\,ds + \end{aligned}$$

$$+\frac{1}{\varepsilon}\int_{\Theta}[\varepsilon f_\varepsilon(\theta,\ x_\varepsilon(s),\ v_\varepsilon(s))-g_\varepsilon(\theta,\ x_\varepsilon(t))]\,q_\varepsilon(d\theta\times ds).$$

And this means (cf. §1) that:

$$\begin{aligned}\frac{d}{ds}\mathsf{M}|v_\varepsilon(s)-\hat{v}_\varepsilon(s)|^2 &\leqslant \frac{-2\gamma}{\varepsilon}\mathsf{M}|v_\varepsilon(s)-\hat{v}_\varepsilon(s)|^2+\\ &+\frac{1}{\varepsilon}\mathsf{M}(\|M(t,\ x_\varepsilon(t))-M(s,\ x_\varepsilon(s))\|\cdot|v_\varepsilon(s)|+\\ &+|a(s,\ x_\varepsilon(s))|+\alpha_\varepsilon|v_\varepsilon(s)-\hat{v}_\varepsilon(s)|)+\\ &+\frac{1}{\varepsilon^2}\int\mathsf{M}|\varepsilon f_\varepsilon(\theta,\ x_\varepsilon(s),\ v_\varepsilon(s))-g_\varepsilon(\theta,\ x_\varepsilon(t))|^2 m_\varepsilon(s,\ d\theta),\end{aligned} \tag{19}$$

where $\alpha_\varepsilon \to 0$. It follows from condition (2) of the theorem that

$$\int\mathsf{M}|\varepsilon f_\varepsilon(\theta,\ x_\varepsilon(s),\ v_\varepsilon(s))-g_\varepsilon(\theta,\ x_\varepsilon(t))|^2 m_\varepsilon(s,\ d\theta)=\beta_\varepsilon,$$

where $\beta_\varepsilon \to 0$. We denote

$$\begin{aligned}\overline{\lim_{\varepsilon\to0}}\sup_{t\leqslant s\leqslant t+h}\mathsf{M}(\|M(t,\ x_\varepsilon(t))-M(s,\ x_\varepsilon(s))\|^2+\\ +\int_{\Theta}|g_\varepsilon(\theta,\ x_\varepsilon(t))-g_\varepsilon(\theta,\ x_\varepsilon(s))|^2 m_\varepsilon(s,\ d\theta))=\delta_\varepsilon(h).\end{aligned}$$

It follows from (17) that

$$\lim_{h\to0}\overline{\lim_{\varepsilon\to0}}\,\delta_\varepsilon(h)=0.$$

Making use of the inequalities

$$\begin{aligned}&\mathsf{M}(\|M(t,\ x_\varepsilon(t))-M(s,\ x_\varepsilon(s))\|\,|v_\varepsilon(s)|+\\ &\quad+|a(s,\ x_\varepsilon(s))|)\,|v_\varepsilon(s)-\hat{v}_\varepsilon(s)|\leqslant\\ &\quad\leqslant\gamma\mathsf{M}|v_\varepsilon(s)-\hat{v}_\varepsilon(s)|^2+\frac{1}{\gamma}\mathsf{M}(\|M(t,\ x_\varepsilon(t))-\\ &\quad-M(s,\ x_\varepsilon(s))\|\,|v_\varepsilon(s)|^2+|a(s,\ x_\varepsilon(s))|^2),\\ &\mathsf{M}\|M(t,\ x_\varepsilon(t))-M(s,\ x_\varepsilon(s))\|^2|v_\varepsilon(s)|^2\leqslant\\ &\quad\leqslant\sqrt{\mathsf{M}\|M(t,\ x_\varepsilon(t))-M(s,\ x_\varepsilon(s))\|^4}\sqrt{\mathsf{M}|v_\varepsilon(s)|^4}\leqslant\\ &\quad\leqslant\sqrt{4\sup_{t,x}\|M(t,\ x)\|^2\cdot\delta_\varepsilon(h)}\sqrt{\mathsf{M}|v_\varepsilon(s)|^4}=O\left(\frac{1}{\varepsilon}\sqrt{\delta_\varepsilon(h)}\right),\end{aligned}$$

we obtain from (19) that for all sufficiently small ε there exists a constant H such that

$$\frac{d}{ds}\mathsf{M}|v_\varepsilon(s)-\hat{v}_\varepsilon(s)|^2\leqslant-\frac{\gamma}{\varepsilon}\mathsf{M}|v_\varepsilon(s)-\hat{v}_\varepsilon(s)|^2+\frac{H}{\varepsilon^2}\sqrt{\delta_\varepsilon(H)}.$$

It follows from this inequality that

$$\mathsf{M}_t |v_\varepsilon(s) - \hat{v}_\varepsilon(s)|^2 \leqslant \frac{H}{\gamma\varepsilon}\sqrt{\delta_\varepsilon(h)}$$

and hence

$$\frac{1}{h}\int_t^{t+h} \varepsilon \mathsf{M} |v_\varepsilon(s) - \hat{v}_\varepsilon(s)|^2 ds = \frac{H}{\gamma}\sqrt{\delta_\varepsilon(h)}.$$

Therefore, to prove (16) it remains to prove that

$$\overline{\lim_{\varepsilon\to 0}} \mathsf{M}\left| \varepsilon \int_t^{t+h} (\hat{v}_\varepsilon(s), z)(\hat{v}_\varepsilon(s), u)\, ds - h(D(t, x_\varepsilon(t))z, u)\right| = o(h). \tag{20}$$

From the equation for $\hat{v}_\varepsilon(s)$ we find that

$$\hat{v}_\varepsilon(s) = \exp\left\{-\frac{s-t}{\varepsilon} M(t, x_\varepsilon(t))\right\} v_\varepsilon(t) + \frac{1}{\varepsilon}\int_t^s \int_\Theta \exp\left\{-\frac{s-\tau}{\varepsilon} M(t, x_\varepsilon(t))\right\} g_\varepsilon(\theta, x_\varepsilon(t))\, q_\varepsilon(d\theta \times d\tau),$$

$$\varepsilon \int_t^{t+h} \left(\hat{v}_\varepsilon(s) - \exp\left\{-\frac{s-t}{\varepsilon} M(t, x_\varepsilon(t))\right\} v_\varepsilon(t), z\right) \times \left(\hat{v}_\varepsilon(s) - \exp\left\{-\frac{s-t}{\varepsilon} M(t, x_\varepsilon(t))\right\} v_\varepsilon(t), u\right) ds =$$

$$= \frac{1}{\varepsilon}\int_t^{t+h} ds \left[\int_t^s \int_\Theta \left(\exp\left\{-\frac{s-\tau}{\varepsilon} M^*(t, x_\varepsilon(t))\right\} z, g_\varepsilon(\theta, x_\varepsilon(t))\right) q_\varepsilon(d\theta \times d\tau) \int_t^s \int_\Theta \left(\exp\left\{-\frac{s-\tau}{\varepsilon} M^*(t, x_\varepsilon(t))\right\} u, g_\varepsilon(\theta, x_\varepsilon(t))\right) q_\varepsilon(d\theta \times d\tau).$$

Since $\mathsf{M}\varepsilon\,|v_\varepsilon(t)|^2$ is bounded and

$$\left\|\exp\left\{-\frac{s-t}{\varepsilon} M^*(t, x_\varepsilon(t))\right\}\right\| \leqslant e^{-\frac{\gamma(s-t)}{\varepsilon}},$$

the expression on the right is equivalent to

$$\varepsilon \int_t^{t+h} (\hat{v}_\varepsilon(s), z)(\hat{v}_\varepsilon(s), u)\, ds$$

being in the mean uniform with respect to t as $\varepsilon \to 0$. On the basis of

Itô's formula, the expression on the right can be rewritten as

$$\frac{1}{\varepsilon}\int_t^{t+h}\int_t^s\int_\Theta\Big(\exp\Big\{-\frac{s-\tau}{\varepsilon}M^*(t,x_\varepsilon(t))\Big\}z,$$
$$g_\varepsilon(\theta,x_\varepsilon(t))\Big)\Big(\exp\Big\{-\frac{s-\tau}{\varepsilon}M^*(t,x_\varepsilon(t))\Big\}u,$$
$$g_\varepsilon(\theta,x_\varepsilon(t))\Big)p_\varepsilon(d\theta\times d\tau)\,ds+$$
$$+\frac{1}{\varepsilon}\int_t^{t+h}\int_\Theta\int_\Theta\iint_{\substack{0<\tau<\tau_1<s\\0<\tau_1<\tau<s}}\Big(\exp\Big\{-\frac{s-\tau}{\varepsilon}M^*(t,x_\varepsilon(t))\Big\}z,$$
$$g_\varepsilon(\theta,x_\varepsilon(t))\Big)\Big(\exp\Big\{-\frac{s-\tau_1}{\varepsilon}M^*(t,x_\varepsilon(t))\Big\}u,$$
$$g_\varepsilon(\theta,x_\varepsilon(t))\Big)q_\varepsilon(d\theta\times dt)\,q_\varepsilon(d\theta_1\times dt_1)\,ds.$$

We denote the factor before $q_\varepsilon(d\theta\times d\tau)q\,(d\theta_1\times d\tau_1)$ under the integral sign by $\psi_\varepsilon(s,\tau,\tau_1,\theta,\theta_1)$. We have that

$$\mathsf{M}\Big[\frac{1}{\varepsilon}\int_t^{t+h}\int_\Theta\int_\Theta\iint_{0<\tau<\tau_1<s}[\psi_\varepsilon(s,\tau,\tau_1,\theta,\theta_1)+$$
$$+\psi_\varepsilon(s,\tau_1,\tau,\theta_1,\theta)]\,q_\varepsilon(d\theta\times d\tau)\,q_\varepsilon(d\theta_1\times d\tau_1)\,ds\Big]^2\leqslant$$
$$\leqslant\frac{2}{\varepsilon^2}\mathsf{M}\int_t^{t+h}\int_\Theta\int_\Theta\int_0^{\tau_1}\Big[\int_{\tau_1}^{t+h}\psi_\varepsilon(s,\tau,\tau_1,\theta,\theta_1)\,ds\Big]^2\times$$
$$\times m_\varepsilon(\tau,d\theta)\,m_\varepsilon(\tau_1,d\theta_1)\,d\tau\,d\tau_1+$$
$$+\frac{2}{\varepsilon^2}\mathsf{M}\int_t^{t+h}\int_\Theta\int_\Theta\int_0^{\tau_1}\Big[\int_{\tau_1}^{t+h}\psi_\varepsilon(s,\tau_1,\tau,\theta_1,\theta)\,ds\Big]^2\times$$
$$\times m_\varepsilon(\tau,d\theta)\,m_\varepsilon(\tau_1,d\theta_1)\,d\tau\,d\tau_1.$$

We shall show that both terms on the right tend to zero. It suffices to consider the first term since the second term is obtained by interchanging z and u. We have that

$$J=\frac{1}{\varepsilon^2}\int_t^{t+h}\int_t^{\tau_1}\int_{\tau_1}^{t+h}\int_{\tau_1}^{t+h}\int_\Theta\int_\Theta\mathsf{M}\psi_\varepsilon(s,\tau,\tau_1,\theta,\theta_1)\times$$
$$\times\psi_\varepsilon(s_1,\tau,\tau_1,\theta,\theta_1)\,ds\,ds_1\times m(\tau,d\theta)\,(m(\tau_1,d\theta_1)\,d\tau\,d\tau_1=$$
$$=\frac{2}{\varepsilon^2}\iiiint_{t<\tau<\tau_1<s<s_1<t+h}\int_\Theta\int_\Theta\mathsf{M}\psi_\varepsilon(s,\tau,\tau_1,\theta,\theta_1)\times$$
$$\times\psi_\varepsilon(s_1,\tau,\tau_1,\theta,\theta_1)\,ds\,ds_1m_\varepsilon(\tau,d\theta)\,m_\varepsilon(\tau_1,d\theta_1)\,d\tau\,d\tau_1.$$

Having made the change of variables $\tau_1-\tau=\varepsilon\sigma_1$, $s-\tau_1=\varepsilon\sigma_2$, $s-s_1=\varepsilon\sigma_3$, $\tau=\tau$ and taking into account that

$$\psi_\varepsilon(s, \tau, \tau_1, \theta, \theta_1) = \hat{\psi}\left(\frac{s-\tau}{\varepsilon}, \frac{s-\tau_1}{\varepsilon}, \theta, \theta_1\right),$$

where

$$\int_0^\infty |\hat{\psi}(\alpha, \beta, \theta, \theta_1)|\, d\alpha\, d\beta \leqslant \frac{|z|\cdot|u|}{\gamma^2} |g_\varepsilon(\theta, x_\varepsilon(t))|\, |g_\varepsilon(\theta_1, x_\varepsilon(t))|,$$

we convince ourselves that $J = O(\varepsilon)$.

We now consider

$$\mathsf{M}\frac{1}{\varepsilon^2}\left(\int_t^{t+h}\int_t^s\int_\Theta \psi_\varepsilon(s, \tau, \tau, \theta, \theta)\, q_\varepsilon(d\theta\times d\tau)\, ds\right)^2 =$$

$$= \mathsf{M}\frac{1}{\varepsilon^2}\int_t^{t+h}\int_\Theta\left[\int_\tau^{t+h}\psi_\varepsilon(s, \tau, \tau, \theta, \theta)\, ds\right]^2 m_\varepsilon(\tau, d\theta)\, d\tau =$$

$$= \mathsf{M}\frac{2}{\varepsilon^2}\iiint\limits_{t<\tau<s<s_1<t+h}\int_\Theta \psi_\varepsilon(s, \tau, \tau, \theta, \theta)\times$$

$$\times\psi_\varepsilon(s_1, \tau, \tau, \theta, \theta)\, m_\varepsilon(\tau, d\theta)\, d\tau\, ds\, ds_1 \leqslant \frac{2}{\gamma^2}(|u|^2 +$$

$$+ |z|^2)\int_t^{t+h}\int_\Theta \mathsf{M}|g_\varepsilon(\theta, x_\varepsilon(t))|^4 m_\varepsilon(\tau, d\theta)\, d\tau \to 0$$

by virtue of condition (2) of the theorem. We have thus established that

$$\lim_{\varepsilon\to 0}\mathsf{M}\Bigg|\varepsilon\int_t^{t+h}(\vartheta_\varepsilon(s), z)(\vartheta_\varepsilon(s), u)\, ds -$$

$$- \frac{1}{\varepsilon}\int_t^{t+h}\int_t^s\int_\Theta\left(\exp\left\{-\frac{s-\tau}{\varepsilon}M^*(t, x_\varepsilon(t))\right\}z, g_\varepsilon(\theta, x_\varepsilon(t))\right)\times$$

$$\times\left(\exp\left\{-\frac{s-\tau}{\varepsilon}M^*(t, x_\varepsilon(t))\right\}u, g_\varepsilon(\theta, x_\varepsilon(t))\right)\times$$

$$\times m_\varepsilon(\tau, d\theta)\, d\tau\, ds\Bigg| = 0.$$

But

$$\frac{1}{\varepsilon}\int_t^{t+h}\int_t^s\int_\Theta\left(\exp\left\{-\frac{s-\tau}{\varepsilon}M^*(t, x_\varepsilon(t))\right\}z, g_\varepsilon(\theta, x_\varepsilon(t))\right)\times$$

$$\times\left(\exp\left\{-\frac{s-\tau}{\varepsilon}M^*(t, x_\varepsilon(t))\right\}u, g_\varepsilon(\theta, x_\varepsilon(t))\right)m_\varepsilon(\tau, d\theta)\, d\tau\, ds =$$

$$= \int_t^{t+h}\int_0^{h/\varepsilon}\int_\Theta(\exp\{-\sigma M^*(t, x_\varepsilon(t))\}z, g_\varepsilon(\theta, x_\varepsilon(t))\times$$

$$\times(\exp\{-\sigma M^*(t, x_\varepsilon(t))\}u, g_\varepsilon(\theta, x_\varepsilon(t)))\, m_\varepsilon(\tau, d\theta)\, d\sigma\, d\tau =$$

$$= \int_t^{t+h} \int_0^{h/\varepsilon} (B(t, x_\varepsilon(t)) \exp\{-\sigma M^*(t, x_\varepsilon(t))\} z,$$

$$\exp\{-\sigma M^*(t, x_\varepsilon(t))\} u)\, d\sigma\, d\tau + o(1) =$$

$$= h \int_0^\infty (B(t, x_\varepsilon(t)) \exp\{-\sigma M^*(t, x_\varepsilon(t))\} z,$$

$$\exp\{-\sigma M^*(t, x_\varepsilon(t))\} u)\, d\sigma + o(1) =$$

$$= h(D(t, x_\varepsilon(t)) z, u) + o(h).$$

The preceding relation shows the validity of (20). Consequently (15) is valid.

8. We now consider martingales in X:

$$\mu_\varepsilon(t) = \int_0^t \int_\Theta h_\varepsilon(\theta, s, x_\varepsilon(s))\, q_\varepsilon(d\theta \times ds),$$

where $h_\varepsilon(\theta, s, x) = M^{-1}(s, x) g_\varepsilon(\theta, x)$. It follows from condition (3) that, for each t, $h > 0$ and $u \in Z$,

$$\lim_{\varepsilon\to 0} |\mathsf{M}[(\mu_\varepsilon(t+h) - \mu_\varepsilon(t), u)^2 - \\ - \int_t^{t+h} (B(s, x_\varepsilon(s)) M^{-1*}(s, x_\varepsilon(s)) u, \\ M^{-1}(s, x_\varepsilon(s)) u)\, ds / \mathcal{F}_t^{(\varepsilon)}]| = 0. \tag{21}$$

We shall show that the set of measures corresponding to the processes $\mu_\varepsilon(t)$ in $D_{[0,T]}(X)$ is compact and that all limiting measures as $\varepsilon \to 0$ are concentrated on $C_{[0,T]}(X)$. To this end, we estimate

$$\mathsf{M}(|\mu_\varepsilon(t+h) - \mu_\varepsilon(t)|^2 / \mathcal{F}_t^{(\varepsilon)}) = \\ = \int_t^{t+h} \mathsf{M}\left(\int_\Theta |h_\varepsilon(\theta, s, x_\varepsilon(s))|^2 m_\varepsilon(s, d\theta) / \mathcal{F}_t^{(\varepsilon)}\right) ds \leqslant c_4 h, \tag{22}$$

where c_4 is some constant. This estimate follows from conditions (3), (4), (5). Hence, the family of measures corresponding to $\mu_\varepsilon(t)$ is compact in $D_{[0,T]}(X)$ (see [11, Vol. 1, p.508]).

Furthermore, on the basis of Itô's formula,

$$\mathsf{M}(\mu_\varepsilon(t+h) - \mu_\varepsilon(t), z)^4 =$$

$$= \mathsf{M} \int_t^{t+h} \int_\Theta 6(\mu_\varepsilon(s) - \mu_\varepsilon(t), z)^2 (h_\varepsilon(\theta, s, x_\varepsilon(s)), z)^2 \times$$

$$\times m_\varepsilon(s, d\theta)\,ds + \mathsf{M}\int_t^{t+h} 4(\mu_\varepsilon(s)-\mu_\varepsilon(t), z)\times$$

$$\times\int_\Theta (h_\varepsilon(\theta, s, x_\varepsilon(s)), z)^3 m_\varepsilon(s, d\theta)\,ds +$$

$$+\mathsf{M}\int_t^{t+h}\int_\Theta (h_\varepsilon(\theta, s, x_\varepsilon(s)), z)^4 m_\varepsilon(s, d\theta)\,ds.$$

It follows from conditions (2) and (3) and the estimate (22) that there exist a constant c_5 and a quantity $\hat{\beta}$ which tends to zero together with ε such that

$$\mathsf{M}\,|\mu_\varepsilon(t+h)-\mu_\varepsilon(t)|^4 \leqslant c_5(h^2+\beta_\varepsilon h).$$

Hence,

$$\overline{\lim_{\varepsilon\to 0}}\,\mathsf{M}\,|\mu_\varepsilon(t+h)-\mu_\varepsilon(t)|^4 \leqslant c_5 h^2.$$

From this, on the basis of Kolmogorov's theorem ([11, Vol. 1, p.237]), we conclude that any limit process will be in the sense of the weak convergence of distributions for the processes $\mu(t)$, a continuous process. We shall assume that the distributions of the two-dimensional processes $\{\mu_\varepsilon(t);\ x_\varepsilon(t)\}$ converges to the distribution of some limit process $\{\bar{\mu}(t),\ \bar{x}(t)\}$. Then, on the basis of (21), for each continuous function $\varphi(x_1, \ldots, x_{2k})$ on X^{2k}, $t_1 < t_2 < \ldots < t_k \leq t$, $h > 0$, $u \in X$,

$$\mathsf{M}\varphi(x(t_1), \ldots, x(t_k), \bar{\mu}(t_1), \ldots, \mu(t_k))\Big[(\bar{\mu}(t+h)-\bar{\mu}(t), u) -$$

$$-\int_t^{t+h}(B(s, x(s))M^{-1*}(s, x(s))u,\ M^{-1*}(s, x(s))u)\,ds\Big]=0.$$

Hence, $(\bar{\mu}(t), u)$ is a continuous martingale with characteristic

$$\int_0^t (M^{-1}(s, x(s))B(s, x(s))M^{-1*}(s, x(s))u,\ u)\,ds.$$

9. Choosing an orthonormalized basis $\{e_1, \ldots, e_r\}$ in X and using formula (15), we can write

$$\varepsilon\int_0^t [M^{-1}(s, x_\varepsilon(s))]'_x(v_\varepsilon(s))v_\varepsilon(s)\,ds =$$

$$= \sum_{k,l=1}^{r} \varepsilon \int_0^t (e_k, v_\varepsilon(s))(e_l, v_\varepsilon(s))[M^{-1}(s, x_\varepsilon(s))]'_x(e_k)e_l ds \sim$$

$$\sim \sum_{j,k,l=1}^{r} \int_0^t (D(s, x_\varepsilon(s))e_k, e_l) \times$$

$$\times ([M^{-1}(s, x_\varepsilon(s))]'_x(e_k)e_l, e_j)ds \cdot e_j.$$

We rewrite relation (12) in the form

$$x_\varepsilon(t) - x_\varepsilon(0) = \int_0^t \tilde{a}(s, x_\varepsilon(s))ds +$$

$$+ \int_0^t \int_\Theta M^{-1}(s, x_\varepsilon(s)) g_\varepsilon(\theta, x_\varepsilon(s)) q_\varepsilon(d\theta \times ds) + \hat{\delta}_\varepsilon(t),$$

where $\hat{\delta}_\varepsilon(t)$ converges to zero in probability ($\tilde{a}$ is defined in equations (7)).

Suppose the distributions $x_{\varepsilon_n}(t)$ converge as $\varepsilon_n \to 0$ to distributions of some process $\bar{x}(t)$. Using the compactness of measures in $D_{[0,T]}(X)$ corresponding to the processes $x_{\varepsilon_n}(t)$, we convince ourselves that the distributions of the processes $x_\varepsilon(t) - x_\varepsilon(0) - \int_0^t \tilde{a}(s, x(s))ds$ will converge to the distributions of the process

$$x_\varepsilon(t) - x_\varepsilon(0) - \int_0^t \tilde{a}(s, x_\varepsilon(s))ds$$

It follows from item 8 that this process will be a continuous martingale and the characteristic of this martingale scalar-multiplied by u is given by the expression (23). Therefore (see Chapter 1),

$$x(t) - x(0) - \int_0^t \tilde{a}(s, x(s)) = \int_0^t \tilde{B}(s, x(s))dw(s),$$

where $\tilde{B}$ is defined by equations (7). Relation (24) is equivalent to (6). Thus, every limit point in the sense of weak convergence for the distribution $x_\varepsilon(t)$ in $D_{[0,T]}(X)$ as $\varepsilon \to 0$ is a measure corresponding to the solution of equation (6). Since, under the conditions of the theorem, this solution is unique, then such a limit point is unique. And now the proof of the theorem follows from the compactness of the family of measures and the uniqueness of the limit point.

1. Baklan, V.V. 'Variational Differential Equations and Markov Processes in Hilbert Space', Dokl. Akad. Nauk SSSR 159 (1964), 707-710 (Russian). MR 30 #1547. English: Soviet Math. 5 (1964), 1553-1556.
2. Bogolyubov, N.N. and Krylov, N.M. 'On the Focker-Planck Equation Derived by a Theory of Perturbation Method, Given on Spectral Properties of a Perturbed Hamiltonian', Zap. kof. mat. fiz. Akad. Nauk Ukr. SSR 4 (1939), 5-158 (Ukrainian).
3. Bogolyubov, N.N. Problems of Dynamical Theory in Statistical Physics. Moscow: Gostehizdat, 1946 (Russian). MR 13 - 196.
4. Bogolyubov, N.N. 'The Equations of Hydrodynamics in Statistical Mechanics', Zb. Prac' Inst. Mat. No. 10 (1948), 41-59 (Ukrainian). MR 14 - 230.
5. Daleckii, Yu. L. 'Infinite-Dimensional Elliptic Operators and the Corresponding Parbolic Equations', Uspehi Mat. Nauk 22, No. 4 (1967), 3-54 (Russian). MR 36 #6821. English: Russian Math. Surveys 22 (1967), 1-53.
6. Dubko, V.A. 'Stochastic Differential Equations in Certain Problems of Mathematical Physics', Dissertation, Kiev, 1979 (Russian).
7. Dynkin, E.B. Markov Processes. Moscow: Fizmatgiz, 1963 (Russian). MR 33 #1886. English edition: New York: Academic Press, 1965. MR 33 #1887.
8. Ersov, M.P. 'Representations of Itô Processes', Teor. Verojatnost. i Primenen. 17 (1972), 167-172 (Russian). MR 45 #9385. English: Theor. Probability Appl. 17 (1972), 165-169.
9. Gihman, I.I. 'On the Influence of a Random Process on a Dynamical System', Nauchn. zap. Kievskogo un-ta 5 (1941), 119-132 (Russian).
10. Gihman, I.I. 'On Passage to the Limit in Dynamical Systems', Nauchn. zap. Kievskogo un-ta 5 (1941), 141-149 (Ukrainian).
11. Gihman, I.I. 'On a Scheme of Formation of Random Processes', Doklady Akad. Nauk SSSR 58 (1947), 961-964 (Russian). MR 9 - 293.
12. Gihman, I.I. 'On the Theory of Differential Equations of Random Processes', Ukrain. Mat. Zurnal 2 (1950), 37-63; 3 (1951), 317-339 (Russian). MR 14 - 61; MR 14 - 1101.
13. Gihman, I.I. and Kadyrova, I.I. 'Some Results of the Investigation of Stochastic Differential Equations', Sb. Teoriya slucainyh proc, I. Kiev: Naukova Dumka, 1973, 51-68. MR 52 #12080. English: in 'Theory of Stochastic Processes, No. 1. New York: John Wiley, 1974.
14. Gihman, I.I. and Skorohod, V.V., Stochastic Differential Equations.

Kiev: Naukova Dumka, 1968 (Russian). MR 41 #7777.

15. Gihman, I.I. and Skorohod, A.V., Theory of Stochastic Processes. Moscow: Nauka, vol. 1 (1971), vol. 2 (1973), vol. 3 (1975) (Russian). MR 49 #6287, MR 49 #6288, MR . English: New York: Springer, 1974. MR 49 #11603.

16. Grigelionis, B. 'A Certain Test of Markovianness for Random Processes', Litovsk. Mat. Sb. 10, No. 2 (1970) 253-258 (Russian). MR 44 #6121.

17. Grigelionis, B. 'The Representation of Integer-Valued Random Measures as Stochastic Integrals over the Poisson Measure', Litovsk. Mat. Sb. 11, No. 11 (1971) 93-101 (Russian). MR 45 #2780.

18. Grigelionis, B. 'The Representation by Stochastic Integrals of Square-Integrable Martingales', Litovsk. Mat. Sb. 14, No. 4 (1974), 53-69, 233-234 (Russian). MR 53 #6734. English: Lithuanian Math. Trans. 14 (1974), 573-584 (1975).

19. Il'in, A.M. and Has'minskii, R.Z., 'On Equations of Brownian Motion', Teor. Verojatnost. i Primenen. 11 (1966), 466-491 (Russian). MR . English: Theor. Probability Appl.

20. Itô, K. 'On a Stochastic Integral Equation', Proc. Japan Acad. 22 (1946), 32-35. MR 12, 191.

21. Itô, K. 'On Stochastic Differential Equations', Mem. Am. Math. Soc. No. 4 (1951). MR 12 - 724.

22. Itô, K. 'Stochastic Differential Equations in a Differential Manifold, I', Nagoya Math. J. 1 (1950), 35-47; II, Mem. Coll. Sci. Univ. Kyoto A28 (1953), 81-85. MR 12 - 425; MR 15 - 636.

23. Itô, K. and Nisio, M., 'On Stationary Solutions of a Stochastic Differential Equation', J. Math. Kyoto Univ. 4 (1964) 1-75. MR 31 #1719.

24. Kolmogorov, A.N. 'On Analytic Methods in Probability Theory', Uspehi Mat. Nauk 5 (1931), 5-41.

25. Krylov, N.V. 'On Quasi-diffusion Processes', Teor. Verojatnost. i Primenen. 11 (1966), 424-443 (Russian). MR . English: Theor. Probability Appl.

26. Krylov, N.V. 'Itô's Stochastic Integral Equations', Teor. Verojatnost. i. Primenen. 14 (1969), 340-348 (Russian). MR 42 #5350. English: Theor. Probability Appl. 14 (1969), 330-336.

27. Lipcer, R.S. and Siryaev, A.N., 'Nonlinear Filtration of Diffusion Markov Processes', Trudy Matem. in-ta im. V.A. Steklova 104 (1968),

135-180 (Russian). MR English: Proc. Steklov Inst. Math.

28. Meier, P.A. Probability and Potentials. Moscow: Mir, 1973 (Russian). MR

29. Portenko, N.I. 'Diffusion Processes with a Generalized Drift Coefficient', Teor. Verojatnost. i. Primenen. 24 (1979), 62-77 (Russian). MR 80g: 600082. English: Theor. Probability Appl.

30. Portenko, N.I. 'Stochastic Differnetial Equations with a Generalized Drift Vector', Teor. Verojatnost. i Primenen. 24 (1979), 332-347 (Russian). MR 81g: 60063. English: Theor. Probability Appl.

31. Skorohod, A.V. Investigations on the Theory of Stochastic Processes. Kiev: Izd-vo Kiev. un-ta, 1961. (Russian). MR

32. Skorohod, A.V. 'On Local Structure of Continuous Markov Processes', Teor. Verojatnost. i Primenen. 11 (1966), 381-425 (Russian). MR 34 #3662. English: Theor. Probability Appl.

33. Skorohod, A.V. 'Stochastic Equations for Quasidiffusion Processes', In the collection: Asimptoticeskie metody teorii nelineinyh kolebanii, Trudy konferencii. Kiev: Naukova Dumka, 1979, pp.211-219 (Russian). MR

34. Skorohod, A.V. 'On the Asymptotic Behvaior of Systems with a Large Number of Randomly Interacting Particles', in the collection: Slucainye processy v zadacah mathematiceskoi fiziki. Kiev: In-t matematiki, 1979, pp.113-138 (Russian). MR

35. Strook, D.W. and Varadhan, S.R.S., 'Diffusion Processes with Continuous Coefficients, I, II', Comm. Pure Appl. Math. 12 (1969), 345-400, 479-530. MR 40 #6641, MR 40 #8130.

36. Uhlenbeck, G.E. and Ford, G.W. Lectures in Statitical Mechanics. Providence, RI: Amer. Math. Soc., 1963. MR 27 #1241. Russian: Moscow, Mir, 1965.

37. Has'minskii, R.Z. Stability of Systems of Differential Equations Under Random Perturbations of their Parameters. Moscow: Nauka, 1969 (Russian). MR 41 #3925.

38. Cantladze, T.L. 'A Stochastic Differential Equation in Hilbert Space', Soobsc. Akad. Nauk Gruzin. SSR 33 (1964), 529-534 (Russian). MR 29 #4094.

39. Shiryaev, A.N. 'Some New Results in the Theory of the Control of Stochastic Processes', Trans. 4th Prague COnfer. Inform. Theory (1965). Prague, 1967, pp.131-203 (Russian). MR

INDEX OF SUBJECTS